Lecture Notes in Computer Science

Edited by G. Goos, J. Hartmanis and J. van Leeuwen

Springer
Berlin
Heidelberg
New York
Barcelona
Hong Kong
London
Milan
Paris
Singapore
Tokyo

Rudolf Eigenmann   Michael J. Voss (Eds.)

OpenMP
Shared Memory
Parallel Programming

International Workshop on OpenMP Applications
and Tools, WOMPAT 2001
West Lafayette, IN, USA, July 30-31, 2001
Proceedings

 Springer

Series Editors

Gerhard Goos, Karlsruhe University, Germany
Juris Hartmanis, Cornell University, NY, USA
Jan van Leeuwen, Utrecht University, The Netherlands

Volume Editors

Rudolf Eigenmann
Michael J. Voss
Purdue University, School of Electrical and Computer Engineering
1285 EE. Bldg., West Lafayette, IN 47907, USA
E-mail: {eigenman/mjvoss}@ecn.purdue.edu

Cataloging-in-Publication Data applied for

Die Deutsche Bibliothek - CIP-Einheitsaufnahme

Conceptual structures : broadening the base ; proceedings / 9th
International Conference on Conceptual Structures, ICCS 2001, Stanford, CA,
USA, July 30 - August 3, 2001. Harry S. Delugach ; Gerd Stumme (ed.). -
Berlin ; Heidelberg ; New York ; Barcelona ; Hong Kong ; London ; Milan ;
Paris ; Singapore ; Tokyo : Springer, 2001
   (Lecture notes in computer science ; Vol. 2120 : Lecture notes in
   artificial intelligence)
   ISBN 3-540-42344-3

CR Subject Classification (1998): C.1-4, D.1-4, F.1-3, G.1-2

ISSN 0302-9743
ISBN 3-540-42346-X Springer-Verlag Berlin Heidelberg New York

Springer-Verlag Berlin Heidelberg New York
a member of BertelsmannSpringer Science+Business Media GmbH

http://www.springer.de

© Springer-Verlag Berlin Heidelberg 2001
Printed in Germany

Typesetting: Camera-ready by author, data conversion by PTP-Berlin, Stefan Sossna
Printed on acid-free paper   SPIN: 10839516   06/3142   5 4 3 2 1 0

Preface

This book contains the presentations given at the Workshop on OpenMP Applications and Tools, WOMPAT 2001. The workshop was held on July 30 and 31, 2001 at Purdue University, West Lafayette, Indiana, USA. It brought together designers, users, and researchers of the OpenMP application programming interface. OpenMP has emerged as the standard for shared memory parallel programming. For the first time, it is possible to write parallel programs that are portable across the majority of shared memory parallel computers. WOMPAT 2001 served as a forum for all those interested in OpenMP and allowed them to meet, share ideas and experiences, and discuss the latest developments of OpenMP and its applications. WOMPAT 2001 was co-sponsored by the OpenMP Architecture Review Board (ARB). It followed a series of workshops on OpenMP, including WOMPAT 2000, EWOMP 2000, and WOMPEI 2000.

For WOMPAT 2001, we solicited papers formally and published them in the form of this book. The authors submitted extended abstracts, which were reviewed by the program committee. All submitted papers were accepted. The authors were asked to prepare a final paper in which they addressed the reviewers comments. The proceedings, in the form of this book, were created in time to be available at the workshop. In this way, we hope to have brought out a timely report of ongoing OpenMP-related research and development efforts as well as ideas for future improvements.

The workshop program included the presentations of the 15 papers in this book, two keynote talks, a panel discussion, and the founding meeting of an OpenMP users' group. The keynote talks were given by David Padua, University of Illinois, entitled "OpenMP and the Evolution of Parallel Programming", and by Larry Meadows, Sun Microsystems, entitled "State of the OpenMP ARB", respectively. The panel was entitled "OpenMP Beyond Shared Memory".

As WOMPAT 2001 was being prepared, the next OpenMP workshop had already been announced, called EWOMP 2001, to be held in Barcelona, Spain. This only adds to the evidence that OpenMP has become a true standard for parallel programming, is very much alive, and is of interest to an increasingly large community.

July 2001   Rudolf Eigenmann

WOMPAT 2001 Program Committee

Tim Mattson, *Intel Corp., USA (Steering Committee Chair)*
Rudolf Eigenmann, *Purdue University, USA (Program Chair)*
Barbara Chapman, *University of Houston, USA (Co-chair)*
Michael Voss, *Purdue University, USA (Co-chair for Local Arrangements)*

Eduard Ayguadé, *Universitat Politecnica de Catalunya, Spain*
Mats Brorsson, *Royal Institute of Technology, Sweden*
Mark Bull, *University of Edinburgh, UK*
Thomas Elken, *SGI, USA*
Larry Meadows, *Sun Microsystems Inc., USA*
Mitsuhisa Sato, *RWCP, Japan*
Sanjiv Shah, *KAI/Intel, USA*

Table of Contents

NUMA Machines and Clusters

OpenMP Extensions

SPEComp: A New Benchmark Suite for Measuring Parallel Computer Performance

Vishal Aslot[1], Max Domeika[2], Rudolf Eigenmann[1], Greg Gaertner[3],
Wesley B. Jones[4], and Bodo Parady[5]

[1] Purdue University,
[2] Intel Corp.,
[3] Compaq Computer Corp.,
[4] Silicon Graphics Inc.,
[5] Sun Microsystems

Abstract. We present a new benchmark suite for parallel computers. SPEComp targets mid-size parallel servers. It includes a number of science/engineering and data processing applications. Parallelism
is expressed in the OpenMP API. The suite includes two data sets,
Medium and Large, of approximately 1.6 and 4 GB in size. Our overview
also describes the organization developing SPEComp, issues in creating
OpenMP parallel benchmarks, the benchmarking methodology underlying SPEComp, and basic performance characteristics.

1 Introduction

Parallel program execution schemes have emerged as a general, widely-used computer systems technology, which is no longer reserved for just supercomputers
and special purpose hardware sytems. Desktop and server platforms offer multithreaded execution modes in today's off-the shelf products. The presence of
parallelism in mainstream computer systems necessitates development of adequate yardsticks for measuring and comparing such platforms in a fair manner.

Currently, no adequate yardsticks exist. Over the past decade, several computer benchmarks have taken aim at parallel machines. The SPLASH [7] benchmarks were used by the research community, but have not been updated recently
to represent current computer applications. Similarly, the Perfect Benchmarks [2]
used to measure high-performance computer systems at the beginning of the
90es. They included standard, sequential programs, which the benchmarker had
to transform for execution on a parallel machine. The Parkbench effort [6] was
an attempt to create a comprehensive parallel benchmark suite at several system levels. However, the effort is no longer ongoing. The SPEChpc suite [4,3] is
a currently maintained benchmark for high-performance computer systems. It
includes large-scale computational applications.

In contrast to these efforts, the goal of the present work is to provide a
benchmark suite that

– is portable across mid-range parallel computer platforms,

R. Eigenmann and M.J. Voss (Eds.): WOMPAT 2001, LNCS 2104, pp. 1–10, 2001.

- can be run with relative ease and moderate resources,
- represents modern parallel computer applications, and
- addresses scientific, industrial, and customer benchmarking needs.

Additional motivation for creating such a benchmark suite was the fact that the OpenMP API (www.openmp.org) has emerged as a de-facto standard for expressing parallel programs. OpenMP naturally offers itself for expressing the parallelism in a portable application suite. The initiative to create such a suite was made under the auspices of the Standard Performance Evaluation Corporation (SPEC, www.spec.org), which has been developing various benchmark suites over the last decade. The new suite is referred to at SPEComp, with the first release being SPEComp2001.

The SPEC organization includes three main groups, the Open Systems Group (OSG), best known for its recent SPEC CPU2000 benchmark, the Graphics Performance Characterization Group (GPC), and the High Performance Group (HPG). The SPEComp initiative was taken by the HPG group, which also develops and maintains the SPEChpc suite. Current members and affiliates of SPEC HPG are Compaq Computer Corp., Fujitsu America, Intel Corp., Sun Microsystems, Silicon Graphics Inc., Argonne National Lab, Leibniz-Rechenzentrum, National Cheng King University, NCSA/University of Illinois, Purdue University, Real World Computing Partnership, the University of Minnesota, Tsukuba Advanced Computing Center, and the University of Tennessee. In contrast to the SPEChpc suite, we wanted to create smaller, more easily portable and executable benchmarks, targeted at mid-range parallel computer systems. Because of the availability of and experience with the SPEC CPU2000 suite, we decided to start with these applications. Where feasible, we converted the codes to parallel form. With one exception, all applications in SPEComp2001 are derived from the CPU2000 suite. We also increased the data set significantly. The first release of the suite includes the Medium data set, which requires a computer system with 2GB of memory. An even larger data set is planned for a future release. Another important difference to the SPEC CPU2000 benchmarks is the run rules, which are discussed in section 2.2.

The remainder of the paper is organized as follows. Section 2 gives an overview of the benchmark applications. Section 3 presents issues we faced in developing the OpenMP benchmark suite. Section 4 discusses basic SPEComp performance characteristics.

2   Overview of the SPEComp Benchmarks

2.1   The SPEComp2001 Suite

SPEComp is fashioned after the SPEC CPU2000 benchmarks. Unlike the SPEC CPU2000 suite, which is split into integer and floating-point applications, SPEComp2001 is partitioned into a Medium and a Large data set.

The Medium data set is for moderate size SMP (Shared-memory MultiProcessor) systems of about 10 CPUs. The Large data set is oriented to systems with

Table 1. Overview of the SPEComp2001 Benchmarks

Benchmark name	Applications	Language	# lines
ammp	Chemistry/biology	C	13500
applu	Fluid dynamics/physics	Fortran	4000
apsi	Air pollution	Fortran	7500
art	Image Recognition/neural networks	C	1300
facerec	Face recognition	Fortran	2400
fma3d	Crash simulation	Fortran	60000
gafort	Genetic algorithm	Fortran	1500
galgel	Fluid dynamics	Fortran	15300
equake	Earthquake modeling	C	1500
mgrid	Multigrid solver	Fortran	500
swim	Shallow water modeling	Fortran	400
wupwise	Quantum chromodynamics	Fortran	2200

30 CPUs or more. The Medium data sets have a maximum memory requirement of 1.6 GB for a single CPU, and the Large data sets require up to 6 GB. Run times tend to be a bit long for people used to running SPEC CPU benchmarks. Single CPU times can exceed 10 hours for a single benchmark on a single state-of-the-art processor. Of the twelve SPEComp2001 applications, nine codes are written in Fortran and three are written in C. Table 1 shows basic features of the benchmarks. The suite includes several large, complex modeling and simulation programs of the type used in many engineering and research organizations. The application areas include chemistry, mechanical engineering, climate modeling, physics, image processing, and decision optimization.

2.2   SPEComp Benchmarking Methodology

The overall objective of the SPEComp benchmark suite is the same as that of most benchmarks: to provide the user community with a tool to perform objective series of tests. The test results serve as a common reference in the evaluation process of computer systems and their components.

SPEComp provides benchmarks in the form of source code, which are compiled according to a specific set of rules. It is expected that a tester can obtain a copy of the suite, install the hardware, compilers, and other software described in another tester's result disclosure, and reproduce the claimed performance (within a small range to allow for run-to-run variation).

We are aware of the importance of optimizations in producing the best system performance. We are also aware that it is sometimes hard to draw an exact line between legitimate optimizations that happen to benefit the SPEComp benchmarks and optimizations that specifically target these benchmarks. However, with the list below, we want to increase awareness among implementors and end users towards the issues related to unwanted benchmark-specific optimizations. Such optimizations would be incompatible with the goal of fair benchmarking.

The goals of the SPEComp suite are to provide a reliable measurement of SMP system performance, and also to provide benchmarks where new technology, pertinent to OpenMP performance, can be evaluated. For these reasons, SPEC allows limited source code modifications, even though it possibly compromises the objectivity of the benchmark results. We will maintain this objectivity by implementing a thorough review process.

To ensure that results are relevant to end-users, we expect that the hardware and software implementations used for running the SPEComp benchmarks adhere to the following conventions:

- Hardware and software used to run the SPEComp benchmarks must provide a suitable environment for running typical C and FORTRAN programs.
- Optimizations must generate correct code for a class of programs, where the class of programs must be larger than a single SPEComp benchmark or SPEComp benchmark suite. This also applies to assertion flags and source code modifications that may be used for peak measurements.
- Optimizations must improve performance for a class of programs where the class of programs must be larger than a single SPEComp benchmark or SPEComp benchmark suite.
- The vendor encourages the implementation for general use.
- The implementation is generally available, documented and supported by the providing vendor.

Benchmarking results may be submitted for a *base*, and optionally, a *peak* run. The base run is compiled and executed with a single set of compiler flags, and no source code modification is permitted. For the peak run, separate compiler flags may be used for each program, and limited source code modifications, restricted to the optimization of parallel performance, are permitted.

3 Issues in Developing an OpenMP Benchmark Suite

Several issues had to be resolved in developing the SPEComp suite. These issues include transforming the original codes to OpenMP, resolving portability problems, defining new data sets, creating self-validation code for each benchmark, and developing benchmark run tool.

3.1 Transforming Sequential Applications to OpenMP

A major effort in creating SPEComp was to convert the original, sequential programs into OpenMP parallel form. We give brief descriptions of the major transformation steps in creating OpenMP parallel programs.

To analyze parallelism in the given codes, we used a mix of application-level knowledge and program-level analysis. In several codes we started by manually inlining subroutine calls for easier interprocedural analysis. We then identified variables that are defined within potential parallel regions, and variables that are "live out" (e.g. formal parameters in a subroutine or function call, function return value, and COMMON blocks).

We declared as PRIVATE scalar variables that are defined in each loop iteration and that are not live out. There were a few instances of LASTPRIVATE variables.

We then identified any scalar and array reductions. Scalar reductions were placed on OpenMP REDUCTION clauses. Array reductions were transformed by hand (Note, that the OpenMP 2.0 specification supports array reductions).

ammp: ammp was by far the most difficult program to parallelize. In addition to directives, we added 16 calls to various OpenMP subroutines. There were only 13 pragmas added, but extensive revisions of the source base were necessary to accommodate parallelism. A few hundred lines of source code were moved or modified to get acceptable scalability. Key to getting scalability was the organization of the program using vector lists instead of linked lists.

applu: We added a total of 50 directives, with almost all simply a form of PARALLEL or DO. A NOWAIT clause was added for the terminator of one loop to improve scalability.

apsi: The main issue in transforming apsi was to privatize arrays that are sections of larger, shared arrays. We did this by declaring separate arrays in the subroutine scope. The size of the arrays is derived from the input parameters (MAX(nx,ny,nz)). Another way to do this is to ALLOCATE the arrays inside OMP PARALLEL but before OMP DO. Several simple induction variable also had to be substituted in this code. In performing these transformations we followed the parallelization scheme of the same code in [5].

art: For art, one difficulty encountered in defining a large data set with reasonable execution time involved the use of the *-objects* switch. When the *-objects N* switch is invoked (N is an integer specifying the number of objects to simulate), the neural network simulates the learning of several more objects than actually trained upon. art's memory requirements scale quadratically with the number of learned objects. Unfortunately, the execution time also scales quadratically due to the F2 layer retry on a mismatch. The art 2 algorithm essentially tests every learned object against the current inputs in a prioritized manner (more probable matches are tested first). To reduce the execution time, the number of F2 layer retries was limited to a small constant value. The reduction still allowed the real objects to be found.

facerec: In this code, pictures in an album are compared against a probe gallery. Both the generation of the album graphs, and the comparison of the album to the probe gallery are parallelized. The parallelization of the probe gallery comparison, which takes up almost all of the computation time, is done on the probe picture level. This is a "shared nothing" parallelization, which is achieved by copying the album graphs into a private array. This method performs well on systems with nonuniform memory access. There still remain many opportunities for parallelism inside the photo gallery comparison code. These opportunities could be exploited using nested parallelism. It would facilitate scaling to larger numbers of processors.

fma3d: fma3d is the largest and most complex code in the suite. There are over 60,000 lines of code, in which we added 127 lines of OpenMP directives. Nearly all directives were of the **PARALLEL** or **DO** variety. There were about a dozen **THREADPRIVATE** directives for a common block, ten reductions, a critical section, and a number of **NOWAIT** clauses. Still, locating the place for these directives was not too difficult and resulted in reasonable scalability.

gafort: We applied three major transformations. First, the "main Generation loop" was restructured so that it is private. It enabled parallelization of the "Shuffle" loop outside this major loop. Inlining of two subroutines expanded this main loop, reducing a large number of function calls. The second transformation concerns the "Random Number Generator". It was parallelized without changing the sequential algorithm. Care was taken to reduce false-sharing among the state variables. Third, the "Shuffle" loop was parallelized using OpenMP locks. The parallel shuffle algorithm differs from the original sequential algorithm. It can lead to different responses for different parallel executions. While this method leads to better scalability and is valid from the application point of view, it made the implementation of the benchmark validation procedure more difficult.

galgel: This code required a bit more attention to parallelization than most, because even some smaller loops in the LAPACK modules need OpenMP directives, and their need became apparent only when a significant number of CPUs were used. A total of 53 directives were added with most being simple **PARALLEL** or **DO** constructs. A total of three **NOWAIT** clauses were added to aid in scalability by permitting work in adjacent loops to be overlapped.

equake: One of the main loops in this code was parallelized at the cost of substantial additional memory allocation. It is an example of memory versus speed tradeoff. Three small loops in the main timestep loop were fused to create a larger loop body and reduce the Fork-Join overhead proportionally. The most time-consuming loop (function smvp) was parallel, but needed the transformation of array reductions. The initialization of the arrays was also parallelized.

mgrid: This code was parallelized using an automatic translator. The code is over 99% parallel. The only manual improvement was to avoid using multiple **OMP PARALLEL/END PARALLEL** constructs for consecutive parallel loops with no serial section in-between.

swim: In this code we added parallel directives to parallelize 8 loops, with a reduction directive needed for one loop. Swim achieves nearly ideal scaling for up to 32 CPUs.

wupwise: The transformation of this code to OpenMP was relatively straightforward. We first added directives to the matrix vector multiply routines for basic OpenMP parallel do in the outer loop. We then added OpenMP directives to the LAPACK routines (`dznrm2.f zaxpy.f zcopy.f zdotc.f zscal.f`). We also inserted a critical section to `dznrm2.f` for a scaling section. Reduction directives were needed for `zdotc.f` and `zscal.f`. After these transformations, wupwise achieved almost perfect scaling on some SMP systems.

3.2   Defining Data Sets and Running Time

An important part of defining a computer benchmark is the selection of appropriate data sets. The benchmark input data has to create an adequate load on the resources of the anticipated test machines, but must also reflect a realistic problem in the benchmark's application domain. In our work, these demands were not easy to reconcile. While we could identify input parameters of most codes that directly affect the execution time and working sets (e.g., time steps and array sizes), it was not always acceptable to modify these parameters individually. Instead, with the help of the benchmark authors, we developed input data sets that correspond to realistic application problems. Compared to the SPEC CPU2000 benchmarks, SPEComp includes significantly larger data sets. This was considered adequate, given the target machine class of parallel servers. The split into a Medium ($< 2GB$) and a Large ($< 8GB$) data set intends to accommodate different machine configurations, but also provides a benchmark that can test machines supporting a 64 bit address space.

3.3   Issues in Benchmark Self-Validation

An important part of a good benchmark is the self validation step. It indicates to the benchmarkers that they have not exploited overly aggressive compiler options, machine features, or code changes. In the process of creating SPEComp we had to resolve several validation issues, which went beyond those arising in sequential benchmarks. One issue is that numerical results tend to become less accurate when computing in parallel. This problem arises not only in the well-understood parallel reductions. We have also observed that advanced compiler optimizations may lead to expression reorderings that invalidate a benchmark run (i.e., the output exceeds the accuracy tolerance set by the benchmark developer) on larger numbers of processors. Another issue arose in benchmarks that use random number generators. The answer of such a program may depend on the number of processors used, making a validation by comparing with the output of a sequential run impossible. To address these issues, we found benchmark-specific solutions, such as using double-precision and identifying features in the program output that are invariant of the number of processors.

3.4   Benchmark Run Tools

Creating a valid benchmark result takes many more steps than just running a code. Procedural mistakes in selecting data sets, applying compilation and execution options, and validating the benchmark can lead to incorrect benchmark reports on otherwise correct programs. Tools that support and automate this process are an essential part of the SPEC benchmarks. The *run tools* for

SPEComp2001 were derived from the SPEC CPU2000 suite. Typically, bench-markers modify only a small configuration file, in which their machine-dependent parameters are defined. The tools then allows one to make (compile and link) the benchmark, run it, and generate a report that is consistent with the run rules described in Section 2.2.

3.5 Portability Across Platforms

All major machine vendors have participated in the development of SPEComp2001. Achieving portability across all involved platforms was an important concern in the development process. The goal was to achieve functional portability as well as performance portability. Functional portability ensured that the makefiles and run tools worked properly on all systems and that the benchmarks ran and validated consistently. To achieve performance portability we accommodated several requests by individual participants to add small code modifications that take advantage of key features of their machines. It was important in these situations to tradeoff machine-specific issues against performance properties that hold generally. An example of such a performance feature is the allocation of lock variables in gafort. The allocation of locks was moved into a parallel region, which leads to better performance on systems that provide non-uniform memory access times.

4 Basic SPEComp Performance Characteristics

We present basic performance characteristics of the SPEComp2001 applications. Note, that these measurements do not represent any SPEC benchmark results. All official SPEC benchmark reports will be posted on SPEC's Web pages (www.spec.org/hpg/omp). The measurements were taken before the final SPEComp2001 suite was approved by SPEC. Minor modifications to the benchmarks are still expected before the final release. Hence, the given numbers represent performance trends only, indicating the runtimes and scalability that users of the benchmarks can expect. As an underlying machine we give a generic platform of which we know clock rate and number of processors.

Figure 1 shows measurements taken on 2, 4 and 8 processors of a 350MHz machine. The numbers indicate the time one has to expect for running the benchmark suite. They also show scalability trends of the suite. Table 2 shows the *parallel coverage* for each code, which is the percentage of serial execution time that is enclosed by a parallel region. This value is very high for all applications, meaning that these codes are thoroughly parallelized. Accordingly, the theoretical "speedup by Amdahl's Law" is near-perfect on 8 processors, as shown in the third column. Column four shows the "Fork-Join" overhead, which is computed as the $(t_{fj} \cdot N)/t_{overall}$, where $t_{overall}$ is the overall execution time of the code, N is the dynamic number of invocations of parallel regions, and t_{fj} is the Fork-Join overhead for a single parallel region. We have chosen $t_{fj} = 10 + p \cdot 2 \ \mu s$, where p is the number of processors. This is a typical value we have observed.

Table 2. Characteristics of the parallel execution of SPEComp2001

Benchmark	Parallel Coverage	Amdahl's Speedup (8 CPU)	% Fork-Join Overhead (8 CPU)	Number of Parallel Sections
ammp	99.2	7.5	0.0008336	7
applu	99.9	7.9	0.0005485	22
apsi	99.9	7.9	0.0019043	24
art	99.5	7.7	0.0000037	3
equake	98.4	7.2	0.0146010	11
facerec	99.9	7.9	0.0000006	$3/2^1$
fma3d	99.5	7.7	0.0052906	$92/30^1$
gafort	99.9	7.9	0.0014555	6
galgel	96.8	6.5	4.7228800	$32/29^1$
mgrid	99.9	7.9	0.1834500	12
swim	99.5	7.7	0.0041672	8
wupwise	99.8	7.9	0.0036620	6

[1] static sections / sections called at runtime

The table shows that the Fork-Join overhead is very small for all benchmarks, except for *galgel*. It indicates that, in all but one codes, the overhead associated with OpenMP constructs is not a factor limiting the scalability. Column five shows the static number of parallel regions for each code. A detailed analysis of the performance of SPEComp2001 can be found in [1].

5   Conclusions

We have presented a new benchmark suite for parallel computers, called SPEComp. We have briefly described the organization developing the suite as well as the development effort itself. Overall, the effort to turn the originally sequential benchmarks into OpenMP parallel codes was modest. All benchmarks are parallelized to a high degree, resulting in good scalability.

SPEComp is the first benchmark suite for modern, parallel servers that is portable across a wide range of platforms. The availability of OpenMP as a portable API was an important enabler. The first release of the suite, SPEComp2001, includes a Medium size data set, requiring a machine with 2 GB of memory. While the codes have been tuned to some degree, many further performance optimizations can be exploited. We expect that the availability of SPEComp will encourage its users to develop and report such optimizations. This will not only lead to improved future releases of the suite, it will also show the value of the new benchmarks as a catalyst for parallel computing technology.

Acknowledgement. We'd like to acknowledge the important contributions of the many authors of the individual benchmark applications. The creation of SPEComp2001 would not have been possible without them.

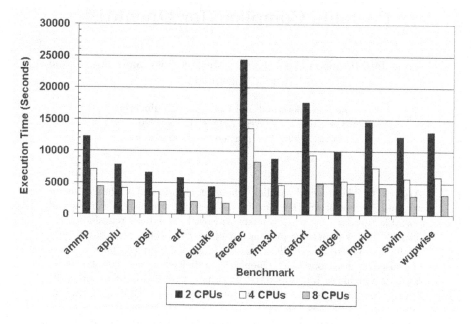

Fig. 1. Execution times of the SPEComp2001 benchmarks on 2,4, and 8 processors of a generic 350MHz machine.

References

1. Vishal Aslot. Performance Characterization of the SPEComp Benchmarks. Master's thesis, Purdue University, 2001.
2. M. Berry, et. al., The Perfect Club Benchmarks: Effective Performance Evaluation of Supercomputers. *Int'l. Journal of Supercomputer Applications*, 3(3):5–40, Fall 1989.
3. Rudolf Eigenmann, Greg Gaertner, Faisal Saied, and Mark Straka. *Performance Evaluation and Benchmarking with Realistic Applications*, chapter SPEC HPG Benchmarks: Performance Evaluation with Large-Scale Science and Engineering Applications, pages 40–48. MIT Press, Cambridge, Mass., 2001.
4. Rudolf Eigenmann and Siamak Hassanzadeh. Benchmarking with real industrial applications: The SPEC High-Performance Group. *IEEE Computational Science & Engineering*, 3(1):18–23, Spring 1996.
5. Rudolf Eigenmann, Jay Hoeflinger, and David Padua. On the Automatic Parallelization of the Perfect Benchmarks. *IEEE Trans. Parallel Distributed Syst.*, 9(1):5–23, January 1998.
6. R. W. Hockney and M. Berry (Editors). PARKBENCH report: Public international benchmarking for parallel computers. *Scientific Programming*, 3(2):101–146, 1994.
7. Steven Cameron Woo, Moriyoshi Ohara, Evan Torrie, Jaswinder Pal Singh, and Anoop Gupta. The SPLASH-2 programs: Characterization and methodological considerations. In *Proceedings of the 22nd International Symposium on Computer Architecture*, pages 24–36, 1995.

Portable Compilers for OpenMP[*]

Seung Jai Min, Seon Wook Kim[**], Michael Voss, Sang Ik Lee, and
Rudolf Eigenmann

School of Electrical and Computer Engineering
Purdue University, West Lafayette, IN 47907-1285
http://www.ece.purdue.edu/ParaMount

Abstract. The recent parallel language standard for shared memory multiprocessor (SMP) machines, OpenMP, promises a simple and portable interface for programmers who wish to exploit parallelism explicitly. In this paper, we present our effort to develop portable compilers for the OpenMP parallel directive language. Our compiler consists of two parts. Part one is an OpenMP parallelizer, which transforms sequential languages into OpenMP. Part two transforms programs written in OpenMP into thread-based form and links with our runtime library. Both compilers are built on the Polaris compiler infrastructure. We present performance measurements showing that our compiler yields results comparable to those of commercial OpenMP compilers. Our infrastructure is freely available with the intent to enable research projects on OpenMP-related language development and compiler techniques.

1 Introduction

Computer systems that offer multiple processors for executing a single application are becoming common place from servers to desktops. While programming interfaces for such machines are still evolving, significant progress has been made with the recent definition of the OpenMP API, which has established itself as a new parallel language standard. In this paper, we present a research compiler infrastructure for experimenting with OpenMP.

Typically, one of two methods is used for developing a shared-memory parallel program: Users may take advantage of a restructuring compiler to parallelize existing uniprocessor programs, or they may use an explicitly parallel language to express the parallelism in the application. OpenMP is relevant for both scenarios. First, OpenMP can be the target language of a parallelizing compiler that transforms standard sequential languages into parallel form. In this way, a parallelizer can transform programs that port to all platforms supporting OpenMP. A first contribution of this paper is to describe such a portable parallelizer. Second, programs written in OpenMP can be compiled for a variety of computer systems, providing portability across these platforms. Our second contribution

[*] This work was supported in part by NSF grants #9703180-CCR and #9872516-EIA.
[**] The author is now with Kuck & Associates Software, A Division of Intel Americas, Inc., Champaign, IL 61820.

is to present such an OpenMP compiler, called PCOMP (Portable Compilers for OpenMP), which is publicly available.

Both compilers are built on the Polaris compiler infrastructure [1], a Fortran program analysis and manipulation system. The public availability of PCOMP and its runtime libraries makes it possible for the research community to conduct such experiments as the study of new OpenMP language constructs, detailed performance analysis of OpenMP runtime libraries, and the study of internal compiler organizations for OpenMP programs. In addition, this paper makes the following specific contributions:

– We describe an OpenMP *postpass* to the Polaris parallelizing compiler, which generates parallel code in OpenMP form.
– We describe a compiler that translates OpenMP parallel programs into thread-based form. The thread-based code can be compiled by a conventional, sequential compiler and linked with our runtime libraries.
– We present performance measurements showing that PCOMP yields results comparable to the commercial OpenMP parallelizers.

We know of two related efforts to provide a portable OpenMP compiler to the research community. The Omni OpenMP compiler [2] translates C and Fortran programs with OpenMP pragmas into C code suitable for compiling with a native compiler linked with the Omni OpenMP runtime library. Also the compiler provides a cluster-enabled OpenMP implementation on a page-based software distributed shared memory. The OdinMP/CCp (C to C with pthreads) [3] is also a portable compiler for C with OpenMP to C with POSIX thread libraries. Both efforts are related to our second contribution.

In Section 2 we present an overview of our OpenMP compiler system. Section 3 and Section 4 describe the details of our OpenMP parallelizer, which generates parallel code with OpenMP directives, and the OpenMP compiler, which translates OpenMP parallel code into thread-based form. In Section 5 we measure the performance of PCOMP and other OpenMP compilers. Section 6 concludes the paper.

2   Portable OpenMP Compilers

Figure 1 gives an overview of our OpenMP compiler system. A program can take two different paths through our OpenMP compiler system: (1) A serial program is analyzed for parallelism by the Polaris Analysis passes and then annotated with OpenMP directives by the OpenMP postpass. The output is an OpenMP parallel program. (2) OpenMP parallel programs can be processed by the OpenMP directive parser and then fed to the MOERAE postpass [4], which transforms them into thread-based code. In this scenario, no additional parallelism is recognized by the Polaris Analysis passes. The generated code can be compiled by a sequential compiler and linked with our MOERAE microtask library.

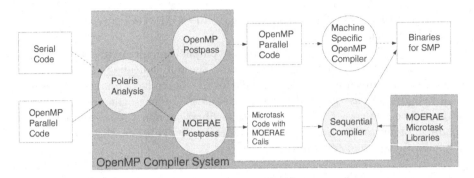

Fig. 1. Overview of our OpenMP compiler system. The same infrastructure can be used to (1) translate sequential programs to OpenMP parallel form and (2) generate code from OpenMP parallel programs. In scenario two, the code is translated into a microtask form, making calls to our MOERAE runtime library.

3   Generating OpenMP from Sequential Program

The Polaris compiler infrastructure is able to detect loop parallelism in sequential Fortran programs. To express this parallelism in OpenMP, we have added an *OpenMP postpass*. It can read the Polaris internal program representation and emit Fortran code annotated with OpenMP directives.

The implementation of this postpass is a relatively straightforward mapping of the internal variables in Polaris to their corresponding directives [5]. Only in the case of array reductions and in privatizing dynamically allocated arrays, more involved language-specific restructuring is done. All other major code restructuring is already implemented in Polaris in a language-independent fashion.

Minor issues arose in that the parallel model used by OpenMP requires some changes to the internal structure of Polaris. Polaris, being tightly coupled to the loop-level parallelism model, lacks a clear method for dealing with parallel regions and parallel sections. Polaris-internally, program analysis information is typically attached to the statement it describes. For example, a loop that is found to be parallel has the DO statement annotated with a *parallel assertion.* The statement is likewise annotated with the variables that are to be classified as shared, private and reduction within the loop nest. In all loop-oriented parallel directive languages that Polaris can generate, this is an acceptable structure.

In order to represent OpenMP parallel regions we have added loop preambles and postambles. Preambles and postambles are code sections at the beginning and end of a loop, respectively, that are executed once by each participating thread. It is now the statement labeled at the beginning of the preamble, which must contain the annotations for parallelism. This statement may no longer be the DO statement. The variables classified as reduction variables, still have to be associated with the DO loop itself. The specification does not require that reductions occur inside loops. You can reduce across parallel regions. These changes in the structure of where information is to be stored, required a review of the Polaris passes.

4   Compiling OpenMP Parallel Programs with PCOMP

PCOMP (Portable Compiler for OpenMP) translates OpenMP parallel programs into thread-based programs. Two capabilities had to be provided to implement PCOMP. One is an extension of the parser for OpenMP directives, and the other is a translator that converts the Polaris-internal representation into the parallel execution model of the underlying machine. As a target we use a thread-based, microtasking model, as is common for loop parallel execution schemes. The Polaris translation pass into thread-based forms has already been implemented in the form of the MOERAE system, described in [4]. MOERAE includes a runtime library that implements the microtasking scheme on top of the portable POSIX-thread libraries.

We added parser capabilities to recognize the OpenMP directives and represent their semantics in the Polaris internal program form. The OpenMP directives are translated in the following manner:

4.1   Parallel Region Construct

The PARALLEL and END PARALLEL directives enclose a parallel region, and define its scope. Code contained within this region will be executed in parallel on all of the participating processors.

Figure 2 illustrates parallel region construct translation. PCOMP *always* replaces these directives with a parallel DO loop whose lower-bound is a constant 1 and upper-bound is the number of participating threads. Figures 2 (a) and (b) show the code transformation from the source to PCOMP intermediate form for a parallel region construct. A variable *mycpuid* indicates the thread identifier number and *cpuvar* denotes the number of participating threads. The newly inserted loop is asserted as PARALLEL, and it allows the MOERAE system to generate the thread-based code shown in Figure 2 (c). At the beginning of the program, the function *initialize_thread* is inserted to initialize the execution environment, such as setting the available number of threads and creating threads. After this initialization, the created threads are in spin-waiting status until function scheduling is called.

The function scheduling invokes the runtime library to manage the threads and copies shared arguments (parameters in Figure 2) into the child threads. Our runtime library provides different variants of this function for the different scheduling options. Currently, only static scheduling is supported.

There are several directive clauses, which may be included on the same line as the PARALLEL directive. They control attributes of the region. The directive clauses are translated as follows.

The IF(*expression*) clause will cause the parallel region to be executed on a single thread, if the expression evaluates to FALSE. This directive will cause a two-version loop to be generated; one loop is serial and the other is parallel.

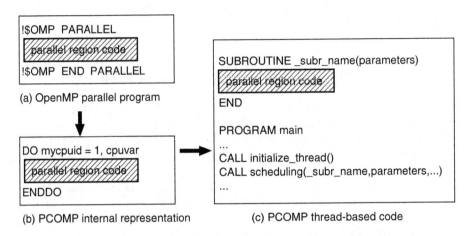

Fig. 2. OpenMP PARALLEL construct transformation. The parallel region is surrounded by a loop whose lower-bound is a constant 1 and upper-bound is the number of participating threads.

The SHARED(*variable_list*) clause includes variables that are to be shared among the participating threads.

The PRIVATE(*variable_list*) clause includes variables that are local to each thread, and for which local instances must exist on each participating thread.

The LASTPRIVATE(*variable_list*) clause is similar to PRIVATE directive. The difference is that when the LASTPRIVATE clause appears on a DO directive, the thread that executes the sequentially last iteration updates the version of the object it had before the construct. PCOMP uses two-version statements to conditionally perform the updates to the variables if the iteration is the last one.

The REDUCTION(*variable_list*) directive clause contains those scalar variables that are involved in a scalar reduction operation within the parallel loop. The Polaris program representation already contains fields with this semantics. PCOMP transforms the OpenMP clauses into these fields. The parallel reduction transformation is then performed using the techniques described in [6]. Preamble and postamble code is generated to initialize a new, private reduction variable and to update the global reduction variable, respectively. The update is performed in a critical section using a lock/unlock pair.

4.2   Work-Sharing Construct

The DO and END DO directives enclose a DO loop inside a region. The iteration space of the enclosed loop is divided among the threads. As an example, the translation process of the DO directive with a REDUCTION clause is described in Figure 3. The PCOMP OpenMP parser reads Work-Sharing directives and assert them to the corresponding DO statement. In Figure 3 (b), there are two DO statements. The first one is a parallel DO loop, which is created by the translation of the PARALLEL construct. The directive clauses, such as PRIVATE and

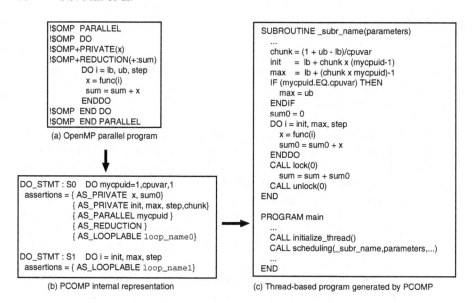

```
!$OMP  PARALLEL
!$OMP  DO
!$OMP+PRIVATE(x)
!$OMP+REDUCTION(+:sum)
        DO i = lb, ub, step
          x = func(i)
          sum = sum + x
        ENDDO
!$OMP  END DO
!$OMP  END PARALLEL
```

(a) OpenMP parallel program

```
SUBROUTINE _subr_name(parameters)
    ...
    chunk = (1 + ub - lb)/cpuvar
    init    = lb + chunk x (mycpuid-1)
    max   = lb + (chunk x mycpuid)-1
    IF (mycpuid.EQ.cpuvar) THEN
        max = ub
    ENDIF
    sum0 = 0
    DO i = init, max, step
      x = func(i)
      sum0 = sum0 + x
    ENDDO
    CALL lock(0)
      sum = sum + sum0
    CALL unlock(0)
END

PROGRAM main
    ...
    CALL initialize_thread()
    CALL scheduling(_subr_name,parameters,...)
    ...
END
```

```
DO_STMT : S0    DO mycpuid=1,cpuvar,1
  assertions = { AS_PRIVATE x, sum0}
              { AS_PRIVATE init, max, step,chunk}
              { AS_PARALLEL mycpuid }
              { AS_REDUCTION }
              { AS_LOOPLABLE loop_name0}

DO_STMT : S1    DO i = init, max, step
  assertions = { AS_LOOPLABLE loop_name1}
```

(b) PCOMP internal representation

(c) Thread-based program generated by PCOMP

Fig. 3. OpenMP PARALLEL REDUCTION DO construct transformation

REDUCTION are transformed to the assertion type information. These assertions give the PCOMP postpass information to generate thread-based code. The second DO statement is the original DO statement whose loop bound indices will be modified to share iteration space by the participating threads.

Figure 3(c) depicts how the iteration space is divided using loop bound modification. The loop bounds are adjusted to reflect the chunk of iterations assigned to each thread. The *if* statement that compares *mycpuid*, a thread number with *cpuvar*, the number of threads detects the last iteration and adjust the upper bound of the loop to come up with the case when the total number of iterations is not a multiple of the number of threads. PCOMP generates postamble code for REDUCTION clauses and encloses the postamble code with lock and unlock functions implemented in the MOERAE runtime libraries [4].

Synchronization Constructs. Critical blocks are surrounded by CRITICAL and END CRITICAL directives. The code enclosed in a CRITICAL/END CRITICAL directive pair will be executed by only one thread at a time. We replace the directives with lock and unlock functions. The BARRIER directive synchronizes all the threads in a team. When encountered, each thread waits until all of the other threads in that team have reached this point. PCOMP replaces the BARRIER directive with a sync function, which is a runtime library function.

The current implementation of PCOMP supports a subset of the OpenMP 1.0 specification for Fortran. Among the unsupported constructs are parallel sections, flushing operation in critical section and dynamic scheduling for parallel loops.

5   Performance Results

We used benchmark programs (WUPWISE, SWIM, MGRID, and APSI) from the
SPEC CPU2000 [7] and the SPEComp2001 benchmark suite to evaluate the
performance of our compiler. We generated executable code using the options f95 -fast -stackvar -mt -nodepend -xvector=no -xtarget=ultra2
-xcache=16/32/1:4096/64/1. The thread-based codes by PCOMP are linked
with the MOERAE runtime libraries. For comparison, we also compiled each
OpenMP code with the SUN Forte 6.1 OpenMP compiler. We ran the codes on
a SUN Enterprise 4000 (Solaris 2.6) system [8] using *ref* data sets.

First, we parallelized the SPEC2000 applications using the Polaris parallelizing compiler [1], which generated the OpenMP codes using our OpenMP postpass. We generated two executable codes: (1) using the SUN parallel compiler,
and (2) using a sequential compiler to compile the translated code by PCOMP
and link with the runtime library. Figure 4 shows the speedup of these codes
relative to the serial execution time of the original codes. The figure shows that
our compiler performs similarly to the commercial compiler. The super-linear
speedup in SWIM is due to a loop interchange in the SHALOW_DO3500 loop by the
parallelizer. In MGRID the performance of our compiler is better than Forte 6.1
in all loops except in RESID_DO600, where our thread-based postpass generates
inefficient code handling LASTPRIVATE variable attributes. Figure 4 shows that
the Polaris-generated codes successfully exploit parallelism in two of the four
benchmarks.

We used an early version of the SPEComp2001 benchmarks for measuring
the performance of our PCOMP compiler. Figure 5 shows the speedup of these
codes relative to the one-processor execution time of the code generated by the
SUN compiler. The figure shows that our compiler performs similarly to the
commercial compiler.

6   Conclusion

We presented an OpenMP compiler system for SMP machines. The compilers
are built using the Polaris infrastructure. The system includes two compilers
for translating sequential programs into OpenMP and for compiling OpenMP
programs in a portable manner (PCOMP), respectively.

We showed that the performance of PCOMP is comparable to commercial
OpenMP compilers. Our infrastructure is publicly available, enabling experimental research on OpenMP-related language and compiler issues. The Polaris
infrastructure has already been widely used in parallelizing compiler research
projects. The availability of an OpenMP compiler component now also supports
this important, emerging standard in parallel programming languages. Currently,
only Fortran77 is supported. Extensions for Fortran90 and C are being developed, providing a complete open-source compiler environment for OpenMP.

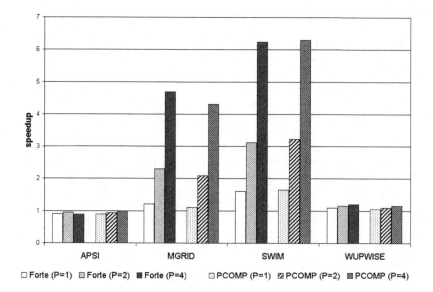

Fig. 4. Automatic generation of OpenMP programs. Speedup of benchmarks as executed on SUN Enterprise 4000 using our OpenMP postpass. The OpenMP codes are parallelized by the Polaris parallelizer with our OpenMP postpass. The codes are compiled by the SUN Forte compiler, and by our PCOMP Portable OpenMP translator with a sequential compiler. P represents the number of processors used.

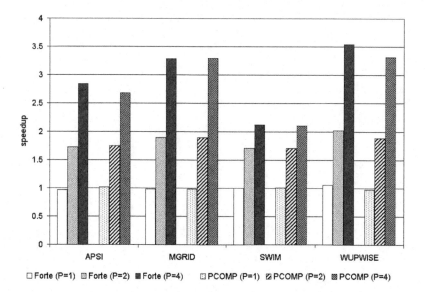

Fig. 5. Compilation of OpenMP programs. Speedup of benchmarks as executed on SUN Enterprise 4000. The OpenMP Suite codes are compiled by the SUN Forte compiler and our PCOMP Portable OpenMP compiler. P represents the number of processors used.

References

1. William Blume, Ramon Doallo, Rudolf Eigenmann, John Grout, Jay Hoeflinger, Thomas Lawrence, Jaejin Lee, David Padua, Yunheung Paek, Bill Pottenger, Lawrence Rauchwerger, and Peng Tu. Parallel programming with Polaris. *IEEE Computer*, pages 78–82, December 1996.
2. Mitsuhisa Sato, Shigehisa Satoh, Kazuhiro Kusano, and Yoshio Tanaka. Design of OpenMP compiler for a SMP cluster. In *The 1st European Workshop on OpenMP (EWOMP'99)*, pages 32–39, September 1999.
3. C. Brunschen and M. Brorsson. OdinMP/CCp - a porable implementation of OpenMP for C. *Concurrency: Practice and Experience*, (12):1193–1203, 2000.
4. Seon Wook Kim, Michael Voss, and Rudolf Eigenmann. Performance analysis of parallel compiler backends on shared-memory multiprocessors. In *Compilers for Parallel Computers (CPC2000)*, pages 305–320, January 2000.
5. Mike Voss. Portable level-parallelism for shared-memory multiprocessor architectures. Master's thesis, Electrical and Computer Engineering, Purdue University, December 1997.
6. Bill Pottenger and Rudolf Eigenmann. Idiom Recognition in the Polaris Parallelizing Compiler. *Proceedings of the 9th ACM International Conference on Supercomputing*, pages 444–448, 95.
7. John L. Henning. SPEC CPU2000: Measuring CPU performance in the new millennium. *IEEE Computer*, July 2000.
8. Sun Microsystems Inc., Mountain View, CA, http://www.sun.com/servers/enterprise/e4000/index.html. *Sun Enterprise 4000*.

The Omni OpenMP Compiler on the Distributed Shared Memory of Cenju-4

Kazuhiro Kusano[1*], Mitsuhisa Sato[1**], Takeo Hosomi[2], and Yoshiki Seo[2]

[1] Real World Computing Partnership, Japan
{kusano,msato}@trc.rwcp.or.jp
[2] C&C Media Research Laboratories, NEC Corporation
{hosomi,seo}@ccm.cl.nec.co.jp

Abstract. This paper describes an implementation and a preliminary evaluation of the Omni OpenMP compiler on a parallel computer Cenju-4. The Cenju-4 is a parallel computer which support hardware distributed shared memory (DSM) system. The shared address space is explicitly allocated at the initialization phase of the program. The Omni converts all global variable declarations into indirect references through the pointers, and generates code to allocate those variables in the shared address space at runtime. The OpenMP programs can execute on a distributed memory machine with hardware DSM by using the Omni. The preliminary results using benchmark programs show that the performance of OpenMP programs didn't scales. While its performance of OpenMP benchmark programs scales poorly, it enables users to execute the same program on both a shared memory machine and a distribute memory machine.

1 Introduction

OpenMP aims to provide portable compiler directives for the shared memory programming environment. The OpenMP directives specify parallel actions explicitly rather than as hints for parallelization. The OpenMP language specification, which is a collection of compiler directives, library routines, and environment variables, came out in 1997 for the Fortran, and in 1998 for the C/C++[1]. While high performance computing programs, especially for scientific computing, are often written in Fortran as the programming language, many programs are written in C in workstation environments. Recently, compiler vendors for PCs and workstations have endorsed the OpenMP API and have released commercial compilers that are able to compile an OpenMP parallel program.

We developed the Omni OpenMP compiler for a shared memory system[2] and for a page-based software distributed shared memory (DSM) system on a cluster of PCs[3]. This paper describes the implementation of the Omni OpenMP compiler on a hardware support DSM system, and examine the performance improvement gained by using the OpenMP programming model.

---

* NEC Corporation, k-kusano@cq.jp.nec.com
** University of Tsukuba, msato@is.tsukuba.ac.jp

R. Eigenmann and M.J. Voss (Eds.): WOMPAT 2001, LNCS 2104, pp. 20–30, 2001.
© Springer-Verlag Berlin Heidelberg 2001

There are many projects related to OpenMP, for example, research to execute an OpenMP program on top of the DSM environment on a network of workstations[10][9], and the investigation of a parallel programming model based on the MPI and the OpenMP to utilize the memory hierarchy of an SMP cluster[8]. Our project also try to develop the OpenMP environment on an SMP cluster, based on the Omni[3].

The remainder of this paper is organized as follows: Section 2 presents the overview of the Omni OpenMP compiler. The hardware support DSM system of the Cenju-4 and the implementation details are presented in section 3. The experimental results and discussion are shown in section 4. The section 5 presents the concluding remarks.

2   The Omni OpenMP Compiler

This section presents the overview of the Omni OpenMP compiler[2][3].

The Omni OpenMP compiler is an open source OpenMP compiler for an SMP machine. [1]. It is a translator which takes an OpenMP program as input and generates a multi-thread C program with run-time library calls. It consists of three parts: the front-ends, the Omni Exc Java toolkit and the run-time library. The front-ends for C and FORTRAN77 are available now. The input programs are translated into an intermediate code, called an Xobject code, by the front-end.

The Exc Java toolkit is a Java class library that provides classes and methods to analyze and transform the Xobject code. It also has function to unparse the Xobject code into a C program. The representation of the Xobject code which is manipulated by the Exc Java toolkit is a kind of Abstract Syntax Tree (AST) with data type information.

The generated C programs are compiled by a native C compiler, and linked with the Omni run-time library to execute in parallel. It uses the POSIX thread library for parallel execution, and this makes it easy to port to other platforms. Platforms that it has already been ported to are the Solaris on the Sparc and on the Intel, the Linux on the Intel, the IRIX on the SGI, and the AIX on the Power.

2.1   Shmem Memory Model

We chose a shmem memory model to port the Omni to a software DSM system[3]. In this model, all global variables are allocated in the shared address space at run time, and those variables are referred through the pointers. The Omni converts all global variable declarations into indirect references through the pointers, and generates code to allocate those variables at run time. This conversion is shown in figure 1 and figure 2. Declaration and reference of global variables in

---

[1] The Omni OpenMP compiler for a SMP platform is freely available at
`http://pdplab.trc.rwcp.or.jp/Omni/`.

```
float x;   /* global variable */
float fx[10];   /* global array */
...
fx[2] = x;
```

Fig. 1. Original code

```
float *__G_x;   /* pointer to x */
float *__G_fx;   /* pointer to fx */
...
*(__G_fx + 2) = (*__G_x);
...
static __G_DATA_INIT(){ /* initialize function */
 _shm_data_init(&(__G_x),4,0);
 _shm_data_init(&(__G_fx),40,0);
}
```

Fig. 2. Converted code

the original code, a variable x and an array fx, are converted to pointers. The allocation of these variables are done in the compiler generated initialization function, '__G_DATA_INIT()' in figure 2. This function is generated for each compilation unit, and called at the beginning of execution. Since the allocation of the DSM is also done by the library call in the program, the shmem memory model can be applied to the hardware support DSM system of the Cenju-4.

2.2   Runtime Library

The runtime library of the Omni on the Cenju-4 is basically the same as the shared memory version, except the machine dependent functions. The barrier is such a function and it simply calls the library function of the Cenju-4. The process of parallel construct is basically the same as the software DSM version Omni. At the start of the 'parallel', pointers to shared variables are copied into the shared address space and passed to slaves, like the software DSM version. The slave processes are waiting to start parallel execution until the trigger from the library occurs. The allocation and deallocation of these threads are managed by using a free list in the runtime library. The list operations are executed exclusively using the system lock function. Nested parallelism is not supported.

3   NEC Cenju-4

The overview of NEC Cenju-4 which is the platform we used is presented in this section.

3.1   Overview

A Cenju-4[4][5] is a distributed memory parallel computer designed and manufactured by the NEC Corporation. Figure 3 shows the overview of the Cenju-4.

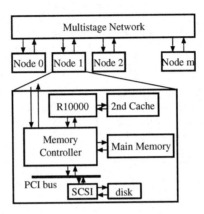

Fig. 3. The Cenju-4 overview

The Cenju-4 is a non-uniform memory access (NUMA) multi-processor. The multistage network of the Cenju-4 connects up to 1024 nodes by using 4 × 4 crossbar switches. The network has the following features: in-order message delivery between any two nodes, multi-cast and gather functions, and a deadlock free mechanism. Each node of the Cenju-4 consists of an R10000 processor, 1M bytes secondary cache, a main memory up to 512M bytes and a controller chip. The controller chip accepts a user level message passing and a DSM access. It is not allowed to use both the message passing and the DSM access on the same page. This attribute is controlled by the operating system that is based on a MACH micro-kernel per page bases.

3.2   Distributed Shared Memory

The DSM of the Cenju-4 is implemented with the use of coherent caches and a directory-based coherence protocol. The DSM has the following four characteristics:

- A directory dynamically switches its representation from a pointer structure to a bit-pattern structure, according to the number of nodes. This scheme requires the constant memory area regardless of the number of processors, and achieves an efficient record of nodes. It achieves scalability in both hardware and performance.
- The Cenju-4 DSM use multicast and gather functions of a network to deliver requests and collect replies. It reduces the overhead of cache invalidation

message. The Cenju-4 also adopts a directory which can specify all nodes caching a block with one memory access.

- A cache coherence protocol that prevents starvation. The Cenju-4 adopts a blocking protocol for cache coherence: requests which cannot be processed immediately are queued in the main memory for later processing. The buffer size of this queue is 32K bytes for 1024 nodes.
- A deadlock-free mechanism with one network. The mechanism that queues certain types of messages for cache coherence in the main memory is provided. The buffer size required for queuing messages is 128K bytes for 1024 nodes. This is allocated in different area from the previous one. This buffer and previous one for starvation are allocated and used in different functions. The cache coherence protocol and the deadlock-free mechanism guarantees shared memory accesses will finish in finite time.

Users have to insert library calls to use the DSM function, since the compiler that can generate codes to utilize the DSM function is not available. The shared address space is allocated by using a library call, and shared variables are allocated or re-allocated in that space. Hereafter, the allocated shared variables can be accessed the same as the private data. These process is almost the same as the Omni on the software DSM system, and the shmem memory model is efficient for the Cenju-4.

There are some restrictions to use the Cenju-4 DSM: first, the maximum size of the shared address space is limited to the physical memory size. Next, the size of the shared address space is limited to 2 Giga bytes. This is because of the MIPS processor architecture that limits user address space up to 2 Giga bytes.

3.3 Execution Model on Cenju-4

The OpenMP uses a fork-join model of a parallel execution. In this model, slave threads are created when the parallel execution is started. A master thread has responsible for the allocation of shared address space and shared variables and setting of parallel execution environment.

The Omni creates slave threads once in a program just after initialization of the master thread. These threads are managed by the run-time library, and remains until the end of a program. The Omni on the software DSM uses this model, but it is implemented on a process on each processor instead of a thread.

The execution model of the Cenju-4 is a MIMD style execution model, like an MPI program. A user program is distributed to all processors by the command 'cjsh', almost the same as 'mpirun' for MPI program. The Omni initialization process has to adapt for this execution model.

An entry function and initialization process is modified as follows: first, all process execute the same entry function, and checks its PE number by using library call. Then, only a master process on the PE0 starts execution of the initialization function, and other processes are waiting to start execution. After the master process ends up the initialization, other slave processes are start to execute the initialization function. Those steps make slave processes possible to share the master process data.

3.4 Shared Address Space

All processes have to execute library 'dsm_open' with the same DSM size as its argument to allocate the shared address space. This library call allocates shared memory for DSM for each processor and enables DSM access. The DSM function is enabled or disabled per page bases by the operating system. The address of shared space can be specified by the user, though the library allocates requested size automatically and returns its address. The size of shared area is limited to 2GB on Cenju-4 DSM.

After the allocation of the shared address space, shared variables are re-allocated in that space explicitly in the program by using 'dsm_malloc' library. This re-allocation is also done by the master thread while its initialization process. This library allocates memory in a block cyclic manner with 128 bytes, the same size as the DSM management block. Each block is dedicated for one variable, even its size is smaller than the 128 bytes.

4 Preliminary Results

This section presents the performance of OpenMP programs on the Cenju-4 DSM.

4.1 Benchmarks and Evaluation Environment

The Cenju-4 with 16 processors is used in the following experiment. Each node has a MIPS R10000 processor with 1M bytes secondary cache and 512M bytes main memory. The multistage network has two stages for 16 PE. The operating system is a native OS of the Cenju-4. The C compiler is based on GNU C compiler and is a cross compiler on a front-end machine, NEC EWS4800. The optimize option '-O3 -funroll-loops' is specified for all experiment. The Cenju-4 system library 'cjsh' is used to invoke a parallel program.

We chose the microbenchmark[6] and the C version NAS Parallel Benchmarks (NPB)[7] which is widely used parallel benchmark programs as our workload. The NPB programs are originally written in Fortran, we rewrote those into C code, and parallelized those using OpenMP directives. The results of CG, FT, IS, and LU are shown in the followings.

4.2 OpenMP Overhead

The microbenchmark[6], developed at the University of Edinburgh, is intended to measure synchronization and loop scheduling overhead of the OpenMP runtime library. The benchmark measures the performance overhead incurred by the OpenMP directives.

We measured the overhead of OpenMP directive by using the microbenchmark. Figure 4 shows the results of 'parallel', 'for', 'parallel-for', 'barrier', and 'parallel-reduction', using the Omni on the Cenju-4 DSM.

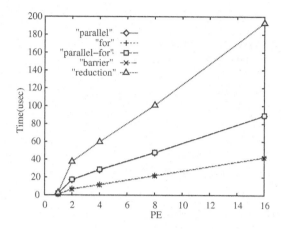

Fig. 4. Omni overhead(usec)

This result shows the overhead is increased in almost linear with the number of processors. The 'barrier' is almost five times slower than the SUN at 4 PE[2]. The overhead of 'parallel' at 4 PE is almost twice as long as the one of the Omni on the SUN 4 PE.

4.3   NPB Parallelization

The NPB programs are parallelized by using OpenMP directives. We use 'orphan' directives to minimize the number of parallel construct and reduces the overhead of parallelization. The code fragment of the parallelized NPB main loop is shown in figure 5.

```
main(){
...
#pragma omp parallel private(it,...)
 for(it=1; it<=NITER; it++){
  <function call, etc>
 }
}
```

Fig. 5. Parallelization of the NPB

The OpenMP 'for' directives are inserted for the parallelizable loops. The scheduling policy for the parallel loops is not specified. It means that the default scheduling of the Omni, static block scheduling, is used. The 'single', 'master' and 'critical' directives are also inserted appropriately. The code fragment of main calculation loop of CG is shown in figure 6.

```
conj_grad(...){
  ...
  for(cgit=...){ ...
#pragma omp for private(k,...)
  for(j=0; j<...){
    for(k=...){
      ...
    }
  }
}
```

Fig. 6. Parallelization of the CG

4.4   NPB Performance Results

The execution time of the NPB programs are shown in the table 1. The 'S' in the table indicates sequential execution. The size of each benchmark program is class A for CG and IS, and class W for FT and LU. The last two programs are class W because of the memory shortage.

Table 1. NPB Execution time(sec)

PE	cg(A)	is(A)	ft(W)	lu(W)
S	62.06	47.83	8.19	386.71
1	73.22	49.90	9.26	396.73
2	60.14	27.15	6.22	241.09
4	38.28	16.14	3.92	125.12
8	22.42	12.85	2.61	68.04
16	11.74	15.28	1.75	37.78

The OpenMP version that is executed on 1 PE is slower than the sequential version from 4% to 14%. This is considered due to the overhead of library calls and inserted codes for parallelization. The performance of CG at 2 PE scales poorly compared to the others.

The speedup that is normalized by sequential execution time is shown in figure 7. The program LU achieves good performance improvement, 10 times speedup at 16 PE. Though the performance of the other programs are improved, the speedup is ranging from 3.1 to 5.3 times at 16 PE. Moreover, the performance of IS is decreased at 16 PE.

4.5   Discussion

The microbenchmark results show that the management policy of the slaves is not appropriate for the Cenju-4. The overhead of the allocation and the deallocation from the list scales poorly.

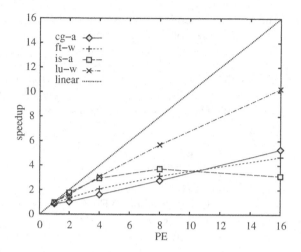

Fig. 7. Speedup of NPB

The results of the NPB benchmark are not so good as the one shown in the previous evaluation[5] that the performance scales well until 32 PE.

The followings are considered as the cause of the difference:

− Parallelization

We parallelize the NPB programs by using the OpenMP directives. The parallelization in the previous evaluation is almost the same way as the one of the MPI version.

− The number of shared variables

The OpenMP attribute of variables, shared or private, are not specified in this experiment. The OpenMP compiler detects global variables and allocated those in the shared address space. On the other hand, the shared variables in the previous experiment are optimized by the user, because it is based on the MPI version. This results the required shared memory area of the OpenMP version is bigger than the one of the previous evaluation.

− The memory alignment of the variables

The shared memory allocation code is generated automatically in the OpenMP version. The allocation of the global data is done by the Cenju-4 library, and it is a cyclic distribution with 128 bytes width. This is the default data distribution of the Cenju-4 DSM library that we used. This allocation has memory fragmentation, because each DSM line is dedicated to only one global variable or array. The alignment of data is depend on the generated sequence of allocation, while the alignment is carefully considered in the previous experiment. This data alignment is expected to affect the result.

− The language of the program and the compiler

The C compiler which is based on the GNU C compiler is used for our experiment. On the other hand, the previous evaluation uses Fortran version

NPB2.3 and the Fortran compiler. This is considered as the main reason of the execution time difference of sequential version.

Although there are undesirable characteristics described above, the parallelization using the OpenMP has a merit. The same programs on the SMP machine can be executed on the DSM system by using our OpenMP compiler. This reduces the program parallelization process of the user in various ways.

5 Concluding Remarks

This paper presented the implementation and the preliminary evaluation of the Omni OpenMP compiler on the Cenju-4. The NEC Cenju-4 is a distributed memory parallel computer and has hardware support DSM function. The OpenMP programs are analyzed and translated to codes that global variables are allocated on the shared address space. The allocation is done in the compiler generated function that is executed in the initialization process. The Omni provides transparent OpenMP environment on the distribute memory system that has hardware support DSM function. This enables users to reduce their task to porting programs. Furthermore, a sequential program and a parallel program are maintained in the same source code.

The result shows the runtime library of the Omni has to be improved to reduce the overhead of parallel construct. The performance of other library functions also has to be checked and to be improved the performance. Our evaluation using the NPB benchmark programs shows that the OpenMP programs achieve speedup on the distributed memory system Cenju-4, though the speedup is not scales.

The Omni now supports data mapping directives for the software DSM system, which enables to schedule the same array element is executed on the same PE. The research of this function on the software DSM system indicated that the data mapping is important to improve performance of the program. The impact of the memory alignment to the application program is one of the main subject of the future work.

References

1. OpenMP Consortium, "OpenMP C and C++ Application Program Interface Ver 1.0", Oct, 1998.
2. K. Kusano, S. Satoh, and M. Sato, "Performance Evaluation of the Omni OpenMP Compiler", WOMPEI, LNCS 1940, pp.403-414, Tokyo, Oct., 2000.
3. M. Sato, H. Harada, and Y. Ishikawa, "OpenMP Compiler for Software Shared Memory System SCASH", WOMPAT 2000, San Diego, July, 2000.
4. Y. Kanoh, T. Hosomi, K. Hirose and T. Nakata, "Design and performance of parallel computer Cenju-4", ISHPC '99, LNCS 1615, pp.55-70, Nara, Sep., 1999.
5. T. Hosomi, Y. Kanoh, M. Nakamura and K. Hirose, "A DSM Architecture for a Parallel Computer Cenju-4", HPCA6, pp.287-298, Toulouse, France, Jan., 2000.

6. J. M. Bull, "Measuring Synchronisation and Scheduling Overheads in OpenMP", EWOMP '99, pp.99-105, Lund, Sep., 1999.
7. David Bailey, E. Barszcz, J. Barton, D. Browning, R. Carter, R. Fatoohi, S. Fineberg, P. Frederickson, T. Lasinski, R. Schreiber, and H. Simon, "The NAS Parallel Benchmarks", RNR-94-007, NAS, 1994.
8. F. Cappello and O. Richard, "Performance characteristics of a network of commodity multiprocessors for the NAS benchmarks using a hybrid memory model", PACT '99, pp.108-116, Oct., 1999.
9. H. Lu, Y. C. Hu and W. Zwaenepoel, "OpenMP on Networks of Workstations", SC'98, Orlando, FL, 1998.
10. Y. C. Hu, H. Lu, A. L. Cox and W. Zwaenepoel, "OpenMP on Networks of SMPs", Proc. of the Thirteenth International Parallel Processing Symposium, pp. 302–310, 1999.

Some Simple OpenMP Optimization Techniques

Matthias Müller

HLRS, University of Stuttgart, D-70550 Stuttgart, Germany,
mueller@hlrs.de,
http://www.hlrs.de/people/mueller

Abstract. The purpose of this benchmark is to test the existence of certain optimization techniques in current OpenMP compilers. Examples are the removal of redundant synchronization constructs and effective constructs for alternative code. The effectiveness of the compiler generated code is measured by comparing different OpenMP constructs and compilers. If possible, we also compare with the hand coded "equivalent" solution.

1 Introduction

Implementations of OpenMP are available on almost every shared memory platform. The portability of applications with OpenMP directives is therefore one of the strong points of this standard. Others are the advantages of shared memory programming in general: it allows a incremental approach to a fully parallel program, and it is also possible to switch between serial and parallel execution during runtime. This enables the programmer to avoid the unavoidable overhead introduced by the parallelism whenever a serial execution is faster.

Since the goal of parallel programming is to achieve higher performance, the further acceptance of OpenMP will strongly depend on compiler optimization techniques, especially in the field where OpenMP has its possible benefits as described above. The importance of benchmarks to measure performance is reflected by many application benchmarks (e.g. [6,7]) and also the well known OpenMP Microbenchmarks [1]. To evaluate the performance of compilers a combination of both has to be used [3]. The focus of this benchmark is somewhat different. On one hand it tries to avoid the architectural dependency of a direct measurement of synchronization and scheduling times, on the other hand it measures isolated OpenMP related optimizations directly without introducing the complex behavior of a complete application. To test whether the proposed optimization techniques are already active in current compilers and to judge the efficiency of the proposed manual solutions several compilers have been used. The compilers from PGI[5], Hitachi and SGI are compilers producing native code, whereas the Omni[4] and guide[2] compiler are front ends to native compilers.

R. Eigenmann and M.J. Voss (Eds.): WOMPAT 2001, LNCS 2104, pp. 31–39, 2001.

2 Benchmarks

The benchmark judges the existence of various optimization techniques by comparing different OpenMP constructs. If a hand coded solutions exists for a construct its performance is also provided.

With one exception all constructs use the same work load:

```
for(n=0; n<count; n++){
  for(i=0;i<length;i++){
    a[i] = b[i]+c[i];
  }
  /* the following is intended to fake the optimizer*/
  b[n]+=offset*a[n];
  c[n]+=offset*a[n];
}
```

The fields a b and c are double precision arrays. The outer loop count is adjusted to guarantee a minimum runtime with a minimum execution count of ten repetitions. Some magic tries to avoid that the optimizer removes the outer loop. Here a zero is added to some array elements. The value of offset is passed in via a command line argument. The inner loop count length is changed to measure the size dependent performance. This is not only important for the constructs that allow alternative code generation for low loop counts, but also provides useful informations for other situations. It shows where the parallelization efforts pays of, and whether or not there is a memory bottleneck. Since only relative performance is important the performance is given in millions loop iterations per second in this paper.

2.1 Overhead of Thread Creation

Whether or not it is beneficial to execute a part of a program in parallel depends on the overhead of thread creation. More precisely it is not the overhead to create the threads for one parallel section, but how the implementation tries to minimize the necessary time to switch between parallel and serial execution in different places. An optimized implementation may decide to keep the threads alive at the end of one parallel section in order to recycle them at the beginning of the next parallel section. The benchmark consists of the standard working loop that sums the two vectors. Three versions of the code are compared against each other:

parallel for: a parallel for construct inside the outer loop. Here, the threads are recreated for every iteration of the outer loop, or at least they have to be recycled.

for: the parallel region is created outside the inner loop. The working loop is parallelized with a for construct. Here, the threads are created only once. This should result in the best performance.

for with barrier: because the `parallel for` also contains an implicit barrier at the end of the construct, the third version contains an explicit barrier in addition to the `for` directive. This checks whether the additional overhead is mainly created by the barrier rather than the thread creation.

Table 1. Overhead of `parallel for` construct, compared to `for` construct with or without barrier.

Compiler	Overhead parallel for [μs]	Overhead for with barrier [μs]	Overhead for [μs]
PC PGI	3.9	4.6	2.8
PC Omni	6.2	3.3	2.4
PC guide	4.7	3.4	2.3
SGI native	11	11	10
SR8K native	2.3	3.7	2.5

The overhead of the PGI compiler is dominated by the implicit barrier at the end of the parallel region. The recycling of threads seems to be very efficient. The thread creation on the Hitachi SR8000 is very fast, there seems to be a very efficient solution for the `parallel for` construct. It is even faster than the `for` construct inside a parallel region. All other compilers show the expected behavior.

2.2   Alternative Code

Once the overhead to create threads is known, the programmer can decide to avoid the overhead of parallel execution for small work load. For this purpose OpenMP contains the `if` clause. A parallel region with the `if` clause is only executed in parallel if the condition is true, otherwise the code is serialized. The goal is to have the performance of the serial code if it is faster than the parallel. To decide whether this goal is achieved the performance of the serial and parallel version is supplied. In addition a manual solution is implemented as follows:

```
    if (condition) {
#pragma omp parallel
        {
            /* code */
        }
    }
    else {
        /* code */
    }
```

In Fig. 1 it is clearly visible that for small loop length the serial code is much faster than the parallel version. A surprise is, that only the native SGI and Hitachi compilers produce code that is as fast as the manual solution. All other compilers are very similar to the Omni OpenMP compiler. Here, the alternative code is better than the simple parallel loop, but inferior to the manual implementation. It should be noted, that the manual implementation requires not more than a simple text replacement. This could easily be implemented by compilers that are front ends to native back end compilers.

2.3   Orphaned Directives

An orphaned directive is a OpenMP directive that does not appear in the lexical extent of a parallel construct, but lies in the dynamic extent. If such a work-sharing construct is not enclosed dynamically within a parallel region the OpenMP standard states "it is treated as though the thread that it encounters it were a team of size one". This allows the user to write code that is used in a parallel and serial context. The question is, whether there will be a overhead if the code is called from a serial context.

Several versions are compared against each other:

scalar: Performance of scalar code.

scalar with function call: Performance of scalar code with function call. The called function contains the working loop. This is done, because many compiler put parallel regions inside functions. This version checks whether a potential performance reduction is caused by the introduced additional call.

orphaned parallel: The working loop contains orphaned OpenMP directives. The loop is called from a parallel region with a team of one thread.

orphaned serial: The working loop contains orphaned OpenMP directives. The loop is called from a serial region.

orphaned manual: Finally, the hand coded version described below of the function is used.

A straight forward manual implementations looks like this:

```
  if (omp_in_parallel()){
#pragma omp for private(i)
    for(i=0;i<length;i++){
      a[i] = b[i]+c[i];
    }
  }
  else {
    for(i=0;i<length;i++){
      a[i] = b[i]+c[i];
    }
  }
```

With the PGI and Hitachi compilers it makes no difference whether the working function is called from a serial region or from a parallel region with a team of one thread. Normally the performance within a parallel region is significantly worse. With the exception of the Hitachi compiler, the manual solution always results in the best performance, although there is an additional call

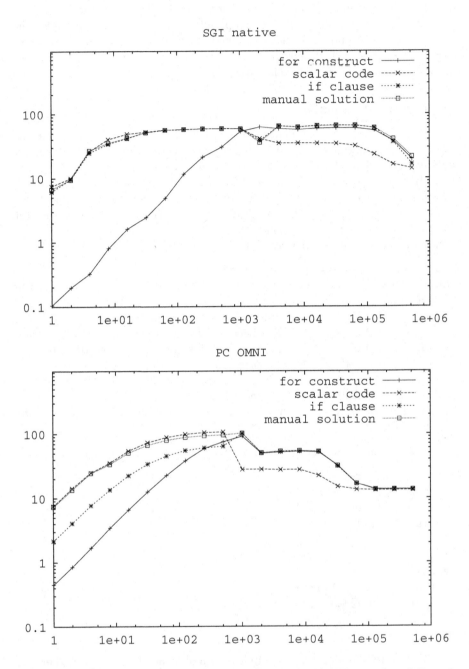

Fig. 1. Alternative code. The codes of the SGI and HITACHI compilers deliver the same performance as the manual solution, all other compilers generate code like the Omni compiler.

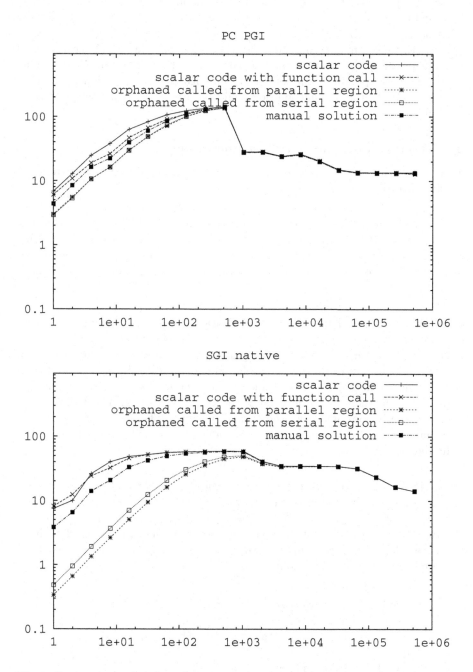

Fig. 2. Performance of orphaned directives. Result of the PGI compiler is on the left, all other compiler produce results similar to the SGI compiler on the right.

to `omp_in_parallel()`. The overhead of orphaned directives is always huge. In principle the compiler could generate code without this overhead. For any function `foo` with orphaned directives, it could generate an additional function `foo_scalar` containing serial code and a function `foo_parallel` with parallel code. Depending whether the function `foo` is called inside or outside a parallel region the corresponding function call is substituted. The function `foo` with a solution equal to the manual solution could be provided to maintain link compatibility.

2.4   Removal of Redundant Synchronization

Three tiny examples check whether the OpenMP compiler removes redundant synchronization.

Parallel Region Merge. This is simply a concatenation of two `parallel for` directives. An optimizing compiler should merge the two parallel regions. All compilers with the exception of the Omni and Hitachi compilers seem to implement this optimization. At least there is no visible performance difference between the two versions.

Implicit `nowait` at End of Parallel Region. The third is a `for` construct directly embedded in a `parallel` region. It is checked whether there is a difference between this version, a direct `parallel for` and a `for` construct with a `nowait` clause. An optimizing compiler should produce the same code for all three versions. Since there is no code between the for loop and the end of the parallel region, there is no need for more than one barrier. The Guide Compiler is the only one that shows the desired behavior. The results of the for loop with `nowait` clause show that the PGI and Omni compiler don't remove the redundant barrier.

Implicit `nowait` Due to Independent Blocks. It consists of two `for` constructs, that work on independent arrays. If the compiler detects that the two basic blocks are independent of each other, it may remove the barrier between the two loops. To increase the difference between a version with and without barrier a load imbalance is introduced in the first loop, that is balanced by a load imbalance in the second loop:

```
for(i=0;i<length;i++){
  /* additional work for first half */
  if ( i < length/2)
     a[i] = b[i]+c[i]+offset*sin(cos(b[i]));
  else
     a[i] = b[i]+c[i];
}
/* same loop on different arrays, additional work on second
                                                      half */
```

The removal of this barrier requires a detailed code analysis. It is no surprise that the front end compilers (Omni and Guide) do not implement this. Also no other compiler performs this optimization.

2.5 Benefit of OpenMP Directives for Sequential Code

This is maybe the most interesting test. The idea behind it is, that OpenMP directives may help the compiler to generate better code because he knows that certain preconditions are fulfilled. As an example workload we use the loop

```
for(i=0; i<size; i++){
  a[index[i]] = a[index[i]] + b[i];
}
```

Because the compiler does not know, how `index[i]` looks like, he cannot assume that the different iterations of the loop are independent. The situation is different, if there is an `#pragma omp parallel for`. If the loop can be executed in parallel it is also subject to optimization techniques like software pipelining or vectorization. For large loop counts the performance of the parallel version executed on one thread should therefore exceed the performance of the serial code. During the tests there was only one case (native compiler on Hitachi SR8000)

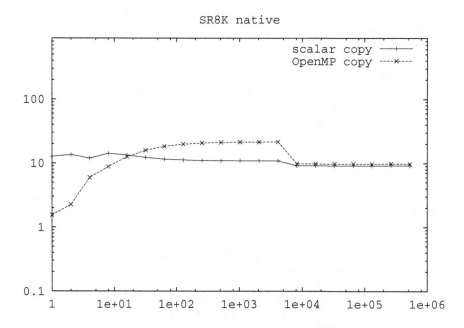

Fig. 3. OpenMP directives may help to optimize serial code. Both versions run with one thread. The OpenMP directive allows to vectorize the code.

where the performance of the loop with OpenMP directives was increased (see Fig. 3). However, a possible further cause might be, that it is impossible to increase the performance of a loop with indirect addressing on the tested architecture.

Table 2. Summary of optimization techniques of different compilers.

Compiler	PGI	Guide	Omni	SGI	Hitachi
Version	3.2-3	3.9	1.2s	7.3.1.1m	?
Optimization					
Alternative code better or equal to manual solution	no	no	no	yes	yes
Optimal Orphaned Directives	no	no	no	no	no
Orphaned Directives better or equal to manual solution	no	no	no	no	yes
Parallel Region Merge	yes	yes	no	yes	no
implicit NOWAIT due to independent blocks	no	no	no	no	no
implicit NOWAIT due to end of region	no	yes	no	no	no
OpenMP directives used as optimization hint	no	no	no	no	yes

3   Conclusion

This small benchmark contains a collection of various optimization techniques that might be implemented in OpenMP compilers. The focus was to avoid architecture dependent techniques on one hand and to concentrate on features that are crucial to achieve maximum performance, especially in areas where the goal is to avoid the parallel overhead whenever a scalar execution is faster.

Five from the seven proposed optimization techniques are already implemented in different compilers, this shows that it is a realistic demand to ask for the implementation of these techniques. Currently most compilers offer only two optimizations, the Hitachi compiler implements three. This demonstrates the possible improvements for the current compilers.

References

1. J. M. Bull. Measuring synchronization and scheduling overheads in OpenMP. In *First European Workshop on OpenMP*, 1999.
2. http://www.kai.com.
3. Kazuhiro Kusano, Shigehisa Satoh, and Mitsuhisa Sato. Performance evaluation of the Omni OpenMP compiler. In *WOMPEI 2000*, Tokyo, Japan, Oct. 2000.
4. Omni OpenMP compiler, http://pdplab.trc.rwcp.or.jp/Omni.
5. http://www.pgroup.com.
6. RWCP. OpenMP version of nas parallel benchmarks.
   http://pdplab.trc.rwcp.or.jp/Omni/benchmarks/NPB/index.html.
7. SPEC. SPEComp 2001. http://www.spec.org.

An Integrated Performance Visualizer for MPI/OpenMP Programs*

Jay Hoeflinger[1], Bob Kuhn[1], Wolfgang Nagel[2], Paul Petersen[1], Hrabri Rajic[1], Sanjiv Shah[1], Jeff Vetter[3], Michael Voss[1], and Renee Woo[1]

[1] KAI Software
Intel Americas, Inc.
Champaign, Illinois 61820
{jay.p.hoeflinger, bob.kuhn, paul.m.petersen, hrabri.rajic, sanjiv.shah, michael.voss, renee.woo}@intel.com
[2] Center for High Performance Computing
Dresden University of Technology, Germany
nagel@zhr.tu-dresden.de
[3] Center for Applied Scientific Computing
Lawrence Livermore National Laboratory
Livermore, California 94551
vetter@llnl.gov

Abstract. As cluster computing has grown, so has its use for large scientific calculations. Recently, many researchers have experimented with using MPI between nodes of a clustered machine and OpenMP within a node, to manage the use of parallel processing. Unfortunately, very few tools are available for doing an integrated analysis of an MPI/OpenMP program. KAI Software, Pallas GmbH and the US Department of Energy have partnered together to build such a tool, VGV. VGV is designed for doing scalable performance analysis - that is, to make the performance analysis process qualitatively the same for small cluster machines as it is for the largest ASCI systems. This paper describes VGV and gives a flavor of how to find performance problems using it.

1 Introduction

Cluster computing has emerged as a defacto standard in parallel computing over the last decade. Now, researchers have begun to use clustered, shared-memory multiprocessors (SMPs) to attack some of the largest and most complex scientific calculations in the world today [8,2], running them on the world's largest machines including the US DOE ASCI platforms: Red, Blue Mountain, Blue Pacific, and White.

MPI has been the predominant programming model for clusters [12]; however, as users move to "wider" SMPs, the combination of MPI and threads has a

---

* This work was performed under the auspices of the U.S. Dept. of Energy by University of California LLNL under contract W-7405-Eng-48.

R. Eigenmann and M.J. Voss (Eds.): WOMPAT 2001, LNCS 2104, pp. 40–52, 2001.

"natural fit" to the underlying system design: use MPI for managing parallelism between SMPs and threads for parallelism within one SMP.

OpenMP is emerging as a leading contender for managing parallelism within an SMP. OpenMP and MPI offer their users very different characteristics. Developed for different memory models, they fill diametrically opposed needs for parallel programming. OpenMP was made for shared memory systems, while MPI was made for distributed memory systems. OpenMP was designed for explicit parallelism and implicit data movement, while MPI was designed for explicit data movement and implicit parallelism. This difference in focus gives the two parallel programming frameworks very different usage characteristics. But these complementary usage characteristics make the two frameworks perfect for handling the two different parallel environments presented by cluster computing: shared memory within a node and distributed memory between the nodes.

Unfortunately, simply writing OpenMP and MPI code does not guarantee efficient use of the underlying cluster hardware. What is more, most existing tools only provide performance information about either MPI or OpenMP, but not both. This lack of integration in our performance tools prevents users from understanding the critical path for performance in their application. To do a good job of performance analysis for such codes, users need detailed information about the expense of operations in their application. Most likely, message passing activity and OpenMP regions are related to the most expensive operations. Viewed in this light, the user needs a performance analyzer to understand the interactions between MPI and OpenMP. For pure message passing codes, several performance analysis tools exist: Vampir [11], TimeScan [3], Paragraph [10], and others. For pure OpenMP codes there is GuideView [7] and a few other proprietary tools from other vendors. For a combination of MPI and OpenMP, we know of only one other tool - Paraver [4].

To address the need for an integrated performance analysis tool, KAI Software and Pallas GmbH have partnered with the Department of Energy through an ASCI Pathforward contract to develop a tool called Vampir/GuideView, or VGV. This tool combines the capabilities of Vampir and GuideView into one tightly-integrated performance analysis tool. From the outset, its design targets performance analysis on systems with thousands of processors.

The purpose of this paper is to describe this tool, how it may be used, and how it can help pin-point the source of performance problems in MPI/OpenMP programs.

2   Related Work

A number of existing tools provide performance analysis of message passing programs. XPVM [5] can be used to analyze the performance of PVM programs. It provides the user an instrumented messaging library and provides a graphical user interface (GUI) for visualizing the performance. The user need not insert instrumentation in their code because the instrumentation exists already in the instrumented library. Vampir [11] and Paragraph [10] are used for MPI programs.

These tools use the MPI profiling interface to capture all MPI calls, then merge the trace information into a single trace file. A visualization program later reads the trace file and draws a graphical representation of the messaging activity between processors.

There are a number of tools for analyzing the performance of HPF codes, which offer a shared memory view to the user, but produce message passing code after having been compiled. The Carnival [13] and Parade [6] systems are examples of these tools. Carnival maintains links to the source code from the instrumentation, so that the user can relate performance to the program, although it was implemented only on IBM systems, to our knowledge. The PARADE system uses by-hand instrumentation and does post-execution "trace animation" through the POLKA animation system.

The Paradyn [1] tool does dynamic instrumentation of a running program by replacing existing instructions with branches to small sections of code called *trampolines* that allow the calling of various instrumentation functions. This provides a very low-overhead and flexible method of instrumentation, but the focus of that project is different from ours since they do not make extensive use of information about the program gathered by a compiler.

The Ovaltine [9] project has developed a tool to analyze the overhead of OpenMP codes, as a way of comparing achieved and achievable performance for a particular code. This type of analysis is already present in the Guideview part of VGV.

The Paraver [4] project, like our own, is focused on building a tool for analyzing the performance of programs that integrate MPI and OpenMP. Paraver is based on a binary instrumenter, that can instrument the MPI functions in a program as well as the OpenMP support functions. Any instrumented region writes trace records to a trace file, which are then displayed through a GUI. The GUI has facilities for user-selected time-scales and zooming in on an arbitrary time range in any display window. The goals of the Paraver project are similar to those of the VGV project, although it is not clear how important scalability is for Paraver. Also, they rely on binary instrumentation, where VGV is based on compiler-inserted instrumentation.

3   Goals of the Project

The main goals of the VGV project are to create an integrated MPI/OpenMP performance analysis tool that is easy to use and that scales well to even the largest systems currently available. This new tool is largely based on the existing Vampir and GuideView tools.

3.1   Scalability

A performance analysis tool faces new problems when it is used for systems with thousands of processors. If the tool is not careful, the amount of information gathered about the performance of a program can become very large, filling disks or causing large data transfer times. The amount of information displayed on-screen can overwhelm the user if it is not displayed appropriately, and on-screen display space is limited, anyway. The aggressive goal of the VGV project is to quadruple the number of processors that can be analyzed every year for the next two years. This year VGV can handle 1000 processors.

3.2 Integration

To perform effective performance analysis with VGV, there must be an integration of information from Vampir and GuideView. This not only avoids the work of manually coordinating output from the two tools, but also provides a platform for synthesizing an overall performance report. The performance data of both tools should also be integrated with source code information.

3.3 Effective Data Presentation

VGV should present an interface which makes the experience of using it for solving performance problems on large machines not materially different from solving such problems on small systems. The tool should also be able to draw the user's attention to potential performance problems, and help the user locate the source of those problems in the program.

4 Using MPI with OpenMP

Before describing how VGV intends to meet its goals, we will briefly mention some key issues that must be addressed when using MPI with OpenMP. MPI may be used with OpenMP, but the two systems have no knowledge of each other, so a few basic rules must be followed to ensure that they do not interfere with each other.

In general, MPI implementations are not thread-safe, so MPI functions can not be safely used when more than one OpenMP thread is active. Therefore, calls to MPI functions should be done either outside OpenMP parallel regions, as shown in Figure 1, or inside a region in which only one thread is active, such as a MASTER region or a SINGLE region, as shown in Figure 2.

In addition, if MPI calls are used in a single-threaded section of an OpenMP parallel region, OpenMP barriers on both sides of the single-threaded section are needed to enforce data consistency. This makes sure that the MPI call sees a consistent view of data, and that the following code section sees any modifications to data caused by the MPI call.

Since message passing calls are hard-coded into the program, the messaging structure of the program can not easily adapt to changing patterns of computation. Therefore, the messaging will typically be done to support a fixed structure in the code. OpenMP, on the other hand, can dynamically adjust the number of threads brought to bear on the various parallel loops within the code, so it can adjust to changes in the fine-grained structure of the computation.

```
        CALL MPI_SEND(A(1), N, MPI_REAL, 1, tag, comm, ierr)
        CALL MPI_RECV(A(1), N, MPI_REAL, 0, tag, comm, status, ierr)
!$omp parallel do shared(A, B, N)
        DO I=1,N
            B(I) = F(A(I))
        END DO
```

Fig. 1. Example of using MPI to exchange data outside an OpenMP parallel region.

```
!$omp parallel shared(A, B, N)
!$omp do
        DO I=1,N
            B(I) = F(A(I))
        END DO
!$omp barrier ! to insure consistent memory
!$omp master
        CALL MPI_ALLREDUCE(A, RA, N, MPI_REAL, MPI_SUM, 0, comm)
!$omp end master
!$omp barrier ! to insure consistent memory
!$omp do
        ...
!$omp end parallel
```

Fig. 2. Example of using MPI to do a reduction operation inside an OpenMP parallel region.

Typically, a fixed number of MPI processes are used, corresponding to the number of nodes being used for the computation. The number of processors within each node would represent the maximum number of processors that can be brought to bear on a parallel loop being run within a single MPI process. In adaptive codes, the amount of work in a particular area of a grid can vary widely, so the number of OpenMP processors used in that area might likewise vary. If there is only a small amount of work in a given parallel loop, then only a small number of processors need be used (less processors used means less synchronization overhead).

5   Structure of the Tool

The flow of the integrated tool follows 4 steps:

1. instrumenting the program at compile time,
2. generating an integrated MPI/OpenMP trace file at runtime,
3. post-run performance analysis for MPI with Vampir,
4. analyzing OpenMP performance with GuideView.

This design integrates Vampirtrace and Vampir with the OpenMP components: Guide, the Guide Runtime Library, and GuideView.

Like most MPI performance analysis tools, Vampirtrace uses the MPI library wrapper interface for instrumentation. As each MPI call is performed, an event is written to a trace file. Vampir is the post-run trace file analysis tool.

Guide is a portable OpenMP compiler for Fortran and C++ that restructures source code and inserts calls to the Guide Runtime Library. The Guide Runtime Library layers on top of threads to implement OpenMP functions. The library is instrumented to call clock timers around all the significant OpenMP events. At the end of a run, the information gathered from these timers is written into a statistics file.

The heart of the MPI and OpenMP integration occurs at runtime. The instrumentation of OpenMP and MPI requires coordination. This is achieved by adding OpenMP events to the Vampirtrace API. The Guide Runtime Library is modified to instrument interesting OpenMP events. For each interesting OpenMP event, the execution times are put into a data structure that is time-stamped and sent to the trace file.

In the next phases of the project, dynamic instrumentation will become more important. Then, the user will be able to identify at run time which parts of the program should be instrumented and traced to get a closer and more focused view to performance bottlenecks. The dynamic instrumentation will combine with and complement the compiler-inserted instrumentation.

6   Usage of the Tool

Once an integrated MPI/OpenMP trace file has been created during the application run, it can be viewed by an integrated user interface. Vampir shows the trace file events ordered by time in the timeline display. When an MPI process executes an OpenMP region, a curvy line glyph appears at the top of that process' time line. The user can select that glyph to view that OpenMP region, or can select a set of MPI processes or a time line section for OpenMP analysis.

OpenMP analysis aggregates the OpenMP data structures from all the trace file events in the selection. Then the aggregated data is written to a file where a GuideView server process reads the file.

GuideView displays the OpenMP regions for each MPI process as a separate set of OpenMP data. In this way, the user can use GuideView tools to select a subset of the hundreds of MPI processes that may be running and sort by any OpenMP performance measure. Examples of OpenMP performance measures for sorting are: scheduling imbalance, lock time, time spent in a locked region, and overhead. The user can specify that GuideView show the top or bottom n, where the user specifies n. This mechanism allows a user to compose compound performance queries by sorting on one criteria, filtering the top responders, and then sorting by another criteria.

The user can also view the subroutine profile for one or a selection of MPI processes within GuideView. This can be viewed as inclusive to allow the user to understand the call tree structure, or exclusive to understand which subroutines consume the most time.

One of the important uses of the tool is to locate regions of the program where some processors spend much time waiting while others are doing useful work. This is referred to as a load-imbalance. From the color-coded display, the user can determine how much time each processor spends waiting.

Besides the new analysis features for the OpenMP parts, the usual analysis features of Vampir can be used for the whole program including all parts. As a new feature, hardware performance monitor information is now available for further inspection. Current processor architectures usually offer performance monitor functionality, mostly in the form of special registers that contain various performance metrics like the number of floating-point operations, cache misses etc. The use of these registers is limited because there is no relation to the program structure: an application programmer typically does not know which parts of the application actually cause bad cache behavior. By extending the Vampir trace format, this data is now available inside the Vampir windows to provide identification mechanisms for functions with low performance properties.

As the systems under investigation could have thousands of processors, the scalability requirement has introduced a couple of new hierarchical concepts for the Vampir windows. Especially, a flexible grouping concept has been developed to show data just related to the right level of abstraction: showing information for all processes, showing just accumulated data for the SMP nodes, showing just information for the master thread, etc. This feature enables end users to easily dive into the important program regions that have performance problems.

A further key use of VGV is source code browsing. The source code associated with any part of the performance data may be brought up in a browsing window by clicking the mouse on the data display.

7   Finding Performance Problems with VGV

Figures 3 and 4 show the performance of the MPI/OpenMP version of the program SWEEP3D. In a hypothetical experiment, a user may have run this program on two MPI processes, with four OpenMP threads in each, and discovered that it exhibits very poor speedup. The user could then run VGV and begin with a whole-program view of the performance, such as the frame at the top of Figure 3. This view shows the execution activity for each MPI process as a horizontal bar. The messaging activity between the processes is shown as lines connecting the two process bars. The regions during which OpenMP activity exists is indicated as regions where the wiggle glyph appears at the top of the process bar. As can be seen from the Figure, nearly all of each process bar is covered by OpenMP parallel regions. So where is the problem?

The problem may be investigated by adding OpenMP detail to the MPI activity. The middle frame in Figure 3 adds OpenMP thread activity to the MPI information. We can begin to advance a theory about the cause of the problem from this frame, because there seem to be a large number of small OpenMP execution regions, separated by large gaps.

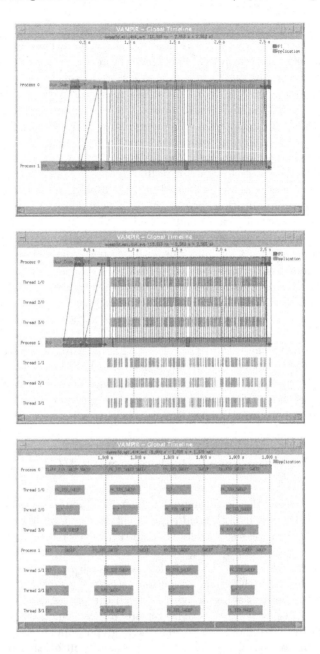

Fig. 3. Performance display frames for the SWEEP3D program, run on two MPI processes with four OpenMP threads on each. The "curvy line glyph" line at the top of a bar indicates that an OpenMP parallel region is active. The top frame shows the overall timeline of the two MPI processes. The middle frame shows the same information, with the OpenMP threads displayed as well. The bottom frame shows the OpenMP thread activity from a single parallel region.

By zooming in on a single OpenMP region (as in the bottom frame of Figure 3), we see that the parallel execution time of each "helper" thread is separated by a gap at least as large as the execution time itself.

This information seems to point to a large number of relatively small parallel regions, dominated by thread startup and shutdown times. To confirm that theory, we can look at aggregated thread information, as in Figure 4. In the top frame of that Figure, we see the aggregated whole-program information for all threads, and the speedup graph for the code. This information corroborates the view that a large fraction of the time spent in each process is "sequential" (the left-most region in each bar). This corresponds to the thread startup and shutdown times. The parallel execution time is indicated in each right-most bar region, and is a small fraction of the total time. The speedup graph shows that the potential for speedup is very poor in this code.

The aggregated performance information for the whole program, displayed per thread (as in the middle frame of Figure 4), confirms this view. All threads are dominated by sequential execution time. Finally, in the bottom frame of Figure 4 we see the thread information displayed for a single OpenMP parallel region, obtained by clicking on the glyph for a single parallel region.

Each VGV display contains implicit links to the original source code for the program generating the performance data. Any of the timeline regions may be "clicked" to obtain a window positioned at the source code that produced the data. By using this feature, it is possible to display the parallel loops within SWEEP3D that we are judging to be "too small". From the source code, we could determine how to increase the amount of work done in the parallel regions.

Comparative analysis can also be done by loading trace files from more than one program run. VGV will plot the results together in any of its frames, to make comparison easy, as in Figure 5. This could allow us to experiment with various input data sets, to see how each affected the performance of the program. In the Figure, three program runs are being compared, the serial version of the program, a 2 (MPI) x 2 (OpenMP) version and a 4 (MPI) x 1 (OpenMP) version.

8   Scalability of the Tool

Prior to this project, GuideView already used light-weight summarization techniques to analyze performance statistics for the OpenMP processors. Vampirtrace, on the other hand, wrote trace records to a single trace file for every MPI call. This produces a potentially very large trace file that must not only be stored, but also completely read and analyzed to provide the user with a display.

To be scalable, VGV must adequately address the following issues:

- the disk storage requirements of an event-based tracing tool could become enormous for long runs with large numbers of processors,
- workstation screens have limited space for displaying performance information,
- simply finding a potential performance problem may be very hard in the blizzard of information potentially generated from a massive run.

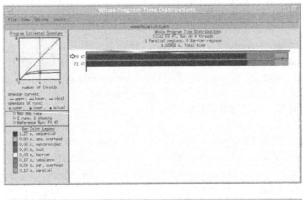

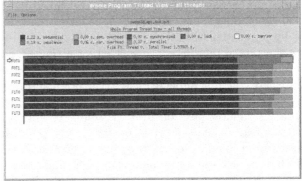

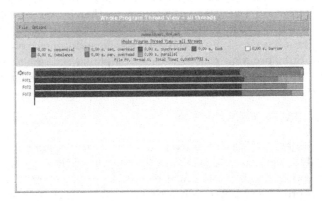

Fig. 4. Performance display frames for the SWEEP3D program, run on two MPI processes with four OpenMP threads on each. The top frame shows the whole program view, displaying aggregated information from all threads and processes. The middle frame shows the threads view of multiple parallel regions. The bottom frame shows the threads view of a single parallel region. In each view, the frames show that the vast majority of time is being spent in sequential execution (the left-most bar region). Parallel execution is the right-most bar region in all cases.

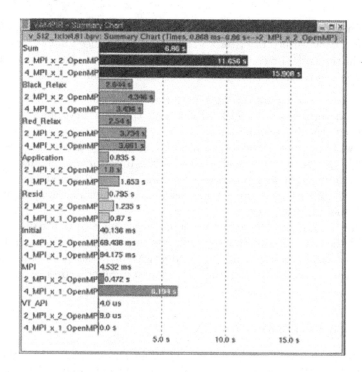

Fig. 5. Comparing three program runs with VGV.

Some of these issues have already been addressed in the current version of VGV. Others will be implemented during the remainder of the project.

VGV will attempt to reduce the size of the trace file through several means:

event compression - Specialized trace records can be used for some events and encoded to save space. Collective communication events, which usually require a full trace record for every process can be reduced to a single trace record and a series of small records, one per task. Also, source code line numbers can be encoded to save space in each trace record.

event combination - Events occurring commonly together can be replaced by a single event. Very short events which are issued until the MPI state changes (e.g. MPI_IProbe / MPI_Test) can be replaced by a single event covering the entire interaction.

event summarization - Some events can be summarized by maintaining only min/max/average values and discarding the events.

structured trace file - The single trace file can be replaced by multiple, hierarchically structured files. This also saves processing time because a top-level summary file can processed much more quickly than can the whole original trace file. This allows the user to see a summary then drill down to other levels in the hierarchy for display.

tracing/instrumentation control - Tracing can be disabled or enabled according to a variety of criteria. The user could place enable/disable trace calls in the code, or could select specific events to enable/disable, or could trace only certain MPI processes, or a variety of other criteria.

The on-screen presentation of the performance information can be made scalable through vertical scrolling of MPI process time-line information, as well as back-to-front stacking of time-lines.

VGV will use data reduction to attempt to identify potential performance problems for the user. Statistical analysis of the data for a single interval of the user's program can identify processes or processors that require unusual (high or low) amounts of various resources (e.g. cache misses, time, memory access time) and mark them for the user.

We have found that the execution time of the analysis tool can be a major fraction of the time required for the tool. For this reason, the tool will be partitioned into a display component (DC) and a trace processing component (TPC). The TPC can be parallelized and run on a small number of processors. The DC can be potentially multi-threaded.

9   Future Directions for VGV

VGV is still in the design and development stage, since we are only in the second year of the three-year project. We expect significant improvements in the system over the remainder of the project.

Possible future directions for VGV include:

- Two way interaction between Vampir and GuideView. When the user selects something in the GuideView display, VGV should map that back to the Vampir Timeline. For example, when selecting a parallel region, a menu item would zoom in on the first instance of the parallel region in the time line so that you could see the actual event distribution inside the parallel region.
- Integrated resource management. Presently, Vampir and GuideView are run in separate processes. This gives them no ability to co-manage the resources they use. If they were combined in the same process, they could coordinate their use of memory, use of the screen and use of disk space. They could also react as a single unit to signals and user requests.
- Move away from Java to some other, more portable, window manager. Java, contrary to popular belief, has proven to be inconsistently portable when it comes to managing the screen display. This is worse on older platforms and can be attributed to some of the early Java implementations. A more mature system, such as Motif, may be more portable.

10   Conclusion

Vampir/GuideView is intended to be a flexible, easy-to-use tool for finding performance problems in programs written with a combination of MPI and

OpenMP, that run for extremely long times and use thousands of processors. We know of no other commercial tool targeted at MPI/OpenMP, and certainly none with the ability to handle the massive runs common on the ASCI machines. During the remaining two years of its development for the ASCI project, we believe that it will become a tool that can be used for the largest ASCI clusters, and will help users quickly pin-point performance anomalies in their codes.

References

1. B. Buck and J.K. Hollingsworth. An API for Runtime Code Patching. *to appear in Journal of Supercomputing Applications and High Performance Computing.*
2. A.C. Calder, B.C. Curtis, and et al. High-Performance Reactive Fluid Flow Simulations Using Adaptive Mesh Refinement on Thousands of Processors. In *Supercomputing 2000: High Performance Networking and Computing Conference*, 2000. electronic pub.
3. Etnus LLC, http://www.etnus.com/Products/TimeScan/index.html. *TimeScan Multiprocess Event Analyzer*, 2001.
4. European Center for Parallelism of Barcelona, Technical University of Catalonia, http://www.cepba.upc.es/paraver/docs/OMPItraceIBM.pdf. *Paraver Reference Manual*, 2000.
5. J.A. Kohl and G.A. Geist. XPVM 1.0 User's Guide. Technical Report ORNL/TM 12981, Oak Ridge National Laboratory, Oak Ridge, Tennessee, November 1996.
6. J.T. Stasko. The PARADE Environment for Visualizing Parallel Program Executions: A Progress Report. Technical Report Technical Report GIT-GVU-95-03, Graphics, Visualization, and Usability Center, Georgia Institute of Technology, Atlanta, GA, January 1995.
7. KAI Software, a division of Intel Americas, http://www.kai.com/parallel/kappro/guideview. *GuideView Performance Analyzer*, 2001.
8. A.A. Mirin, R.H. Cohen, and et al. Very High Resolution Simulation of Compressible Turbulence on the IBM-SP System. In *Supercomputing '99: High Performance Networking and Computing Conference*, 1999. electronic pub.
9. M.K. Bane and G.D. Riley. Automatic Overheads Profiler for OpenMP Codes. In *Proceedings of the European Workshop on OpenMP (EWOMP) 2000, Edinburgh, Scotland, U.K.*, September, 2000.
10. M.T. Heath and J.A. Etheridge. Visualizing the Performance of Parallel Programs. *IEEE Software*, 8(5):29–39, September 1991.
11. Pallas GmbH, http://www.pallas.de/pages/vampir.htm. *Vampir 2.5 - Visualization and Analysis of MPI Programs*, 2001.
12. G.F. Pfister. *In Search of Clusters: The Coming Battle in Lowly Parallel Computing.* Prentice Hall, Upper Saddle River, NJ, 1995.
13. W. Meira Jr. and T.J. LeBlanc and A. Poulos. Waiting Time Analysis and Performance Visualization in Carnival. In *ACM SIGMETRICS Symp. on Parallel and Distributed Tools*, May 1996.

A Dynamic Tracing Mechanism for Performance Analysis of OpenMP Applications

Jordi Caubet[1], Judit Gimenez[1], Jesus Labarta[*][1], Luiz DeRose[2], and Jeffrey Vetter[**][3]

[1] European Center for Parallelism of Barcelona
Department of Computer Architecture
Technical University of Catalonia
Barcelona, Spain
{jordics, judit, jesus}@cepba.upc.es
[2] Advanced Computing Technology Center
IBM T. J. Watson Research Center
Yorktown Heights, NY, USA
laderose@us.ibm.com
[3] Center for Applied Scientific Computing
Lawrence Livermore National Laboratory
Livermore, CA, USA
vetter@llnl.gov

Abstract. In this paper we present OMPtrace, a dynamic tracing mechanism that combines traditional tracing with dynamic instrumentation and access to hardware performance counters to create a powerful tool for performance analysis and optimization of OpenMP applications. Performance data collected with OMPtrace is used as input to the Paraver visualization tool for detailed analysis of the parallel behavior of the application. We demonstrate the usefulness of OMPtrace and the power of Paraver for tuning OpenMP applications with a case study running the US DOE ASCI Sweep3D benchmark on the IBM SP system at the Lawrence Livermore National Laboratory.

1 Introduction

OpenMP has emerged as the standard for shared memory parallel programming, allowing users to write applications that are portable across most shared memory multiprocessors. However, in order to achieve high performance on these systems, application developers still face a large number of application performance problems, such as load imbalance and false sharing. These performance

[*] This work was partially supported by IBM under a Shared University Research grant and by the Spanish Ministry of Education (CICYT) under contract TIC98-0511
[**] This work was performed under the auspices of the U.S. Dept. of Energy by University of California LLNL under contract W-7405-Eng-48. LLNL Document Number UCRL-JC-142770.

R. Eigenmann and M.J. Voss (Eds.): WOMPAT 2001, LNCS 2104, pp. 53–67, 2001.
© Springer-Verlag Berlin Heidelberg 2001

problems make application tuning complex and often counter-intuitive. Moreover, these problems are hard to detect without the help of performance tools that have low intrusion cost and are able to correlate dynamic performance data from both software and hardware measurements.

In this paper, we describe OMPtrace - a dynamic tracing mechanism that combines traditional tracing with dynamic instrumentation and access to hardware performance counters - to create a powerful tool for performance analysis and optimization of OpenMP applications.

OMPtrace is built on top of the Dynamic Probe Class Library (DPCL)[2], an object-based C++ class library and runtime infrastructure that flexibly supports the generation of arbitrary instrumentation, without requiring access to the source code. DPCL allows the instrumentation of the OpenMP runtime systems, providing the flexibility to measure the overhead of initialization and finalization of parallel regions. For detailed analysis of the parallel behavior of the application, OMPtrace data is, then, analyzed with the Paraver visualization tool[3]. To demonstrate the power of OMPtrace and Paraver, we analyze the performance of the Sweep3D application[4] as a case study.

The remainder of this paper is organized as follows. In Section 2 we briefly describe the main DPCL issues. In Section 3 we discuss the OMPtrace interface. In Section 4 we present a case study where we describe the steps followed to analyze the Sweep3D application and how OMPtrace and Paraver were useful to identify potential improvements. Finally, our conclusions are summarized in Section 5.

2   The Dynamic Probe Class Library

Traditionally, instrumentation systems have had to strike a balance between minimizing instrumentation overhead and maximizing the amount of performance data captured. One approach to managing instrumentation overhead is to limit both the number of events recorded and the size of those events. However, this could mean that key events may not have been recorded. Likewise, if too much instrumentation is inserted, the overhead may be so high that it is no longer representative of the un-instrumented program's execution behavior.

Another challenge is that many instrumentation systems require that programs be re-compiled after being instrumented. While this is generally possible, for large applications it can be time consuming. Even worse, for third party libraries and applications users where the source code may not be available, re-compiling will not be possible. An alternative is to allow a program to be modified while it is executing, and thereby eliminate the need to re-compile, re-link, or re-execute the program.

Dynamic instrumentation provides the flexibility for tools to insert probes into applications only where it is needed. The Dynamic Probe Class Library, developed at IBM, is an extension of the dynamic instrumentation approach, pioneered by the Paradyn group at the University of Wisconsin[5]. DPCL is built on top of the Dyninst Application Program Interface (API)[1]. Using DPCL, a performance tool can attach to an application, insert code patches into the binary and start or continue its execution. Access to the source code of the target application is not required and the program being modified does not need to be re-compiled, re-linked, or even re-started.

DPCL provides a set of C++ classes that allows tools to connect, examine, and instrument a spectrum of applications: single processes to large parallel applications. DPCL is composed of a client library, a runtime library, a daemon, and a super-daemon. End user tools can be created with the client library. The runtime library supports instrumentation generation and communication. The daemon interfaces with the Dyninst library to instrument and manage user processes; a super-daemon manages security and client connections to these DPCL daemons. With DPCL, program instrumentation can be done at function entry points, exit points, and call sites.

3   The OMPtrace

The integration of DPCL into OMPtrace is based on the fact that the IBM compiler translates OpenMP directives into function calls. Figure 1 shows, as an example, the compiler transformations for an OpenMP parallel loop. The OpenMP directive is translated into a call to a function from the OpenMP runtime library ("xlsmpParDoSetup"), which is responsible for thread management, work distribution, and thread synchronization. The loop is transformed into a function ("A@OL1" in the example in Figure1) that is called by each of the OpenMP threads.

Since DPCL allows the installation of probes at function call entry and exit points, as well as before and after a function call, the OMPtrace tool installs two pairs of probes for each parallel region in the target application. As shown in Figure 2 the first pair (DPCL probe (1)) is inserted before and after the call to the OpenMP runtime library function, while the second pair (DPCL probe(2)) is inserted at the call entry and exit point of the parallel region. Given these two pairs of probes, one can measure the overhead of starting and terminating a parallel region. Additionally, a third pair of probes (DPCL probe (3)) is inserted at the call entry and exit points of each function that contains an OpenMP parallel region.

Figure 3 displays the startup procedure executed by OMPtrace. The tool communicates with the DPCL daemon (1), which in turn acts on the application binary (2). Probe installation (3) is executed in two steps. First, OMPtrace requests the DPCL communication daemon to load the tracing module into the target application. This module contains the functions that will be called by the probes. Once the tracing module has been loaded, OMPtrace requests the communication daemon to inserts the probes into the application. After the probes are installed, OMPtrace starts the application. Notice that nothing precludes OMPtrace from attaching to a running application and execute the same procedure. We are considering this feature for future work.

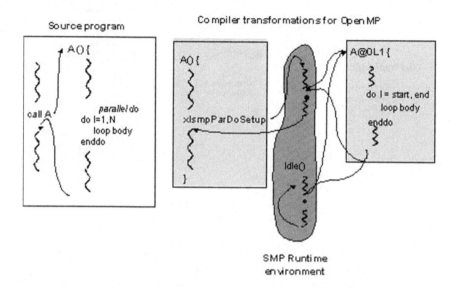

Fig. 1. Compiler transformations for an OpenMP parallel loop.

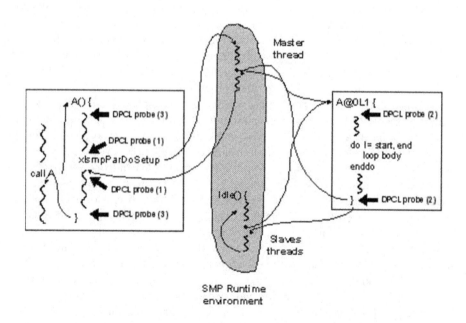

Fig. 2. DPCL probes on functions that contain parallel regions.

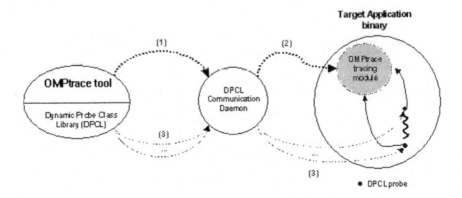

Fig. 3. Application startup with OMPtrace.

Similarly, OMPtrace can be used to instrument locks that are used to ensure mutual exclusion in the application. In this case, dynamic instrumentation is placed before and after the OpenMP functions that are called to handle the locks, generating an event trace every time that a thread enters in one of the following four states: trying to acquire a lock, lock acquisition, starting to release a lock, and lock release. This instrumentation is useful to measure lock overhead, contention in critical sections, and the actual pattern of lock acquisition. Since this instrumentation may introduce a significant overhead, especially for very small critical sections, it is only activated when specified by the user with command line flags when executing OMPtrace. It is our experience that even in cases where the overhead is significant, the information on the pattern and interaction between threads that this tracing facility provides is very helpful to improve the performance of the parallel program.

Another OMPtrace feature is the ability to automatically access hardware counters. The IBM Power3 processor provides 8 counters, each one able to count a number of hardware events. OMPtrace allows users to select any valid combination of hardware events, via an environment variable. By default, OMPtrace uses a standard set of events to count instructions, floating point operations, fused multiply adds (FMAs), and loads. When the hardware counters option is activated, OMPtrace emits event records at the entry and exit of every instrumented point in the program, identifying the hardware events being collected and for each event, the count between the current and the previous tracing point. In order to avoid excessive overhead and reduce trace file size for the default analysis, this feature is also only activated via a command line flag.

One of the known weakness of hardware performance counters is that they only provide raw counts, which does not necessarily help users to identify which events are responsible for bottlenecks in the program performance. However, Paraver has a very flexible mechanism to compute and display a large number of performance indices and derived metrics from the information emitted into the trace by OMPtrace. Thus, the hardware counter information included in the

trace file can be later processed by Paraver to generate a large number of performance indices, which allows users to correlate the behavior of the application to
one or more of the components of the hardware. For example, Paraver can display as a function of time for a given routine (or interval) the quotient between
the number of L1 misses (as reported by the event at the exit of the routine)
and the duration of the routine. Indices such as L1 misses per second or floating
point operations per second can be visualized as a function of time. Additionally, a second level of semantic functions can be obtained by combining (i.e.,
adding, dividing, etc.) the functions of time computed directly from the records
in the trace as stated above. We call this feature *derived windows*. For example,
starting with a window that looks at *"cycles"* to compute the number of cycles
for each function (or interval) and other window that looks at the *"instructions"*
it is possible to derive an IPC window by dividing those two windows. This
derived window will display the actual Instructions per Cycle obtained for each
interval of the application, which can be useful for example to compare with the
theoretical limit of the machine (4 issue in the Power3 case).

An interesting way to use these derived windows is to build performance
models of the processor and try to explain the performance of the application
based on these models. For example, one can compute the theoretical IPC limit
considering just the number of floating point operations and the number of
misses by taking into account that only two FPUs are available and assuming a
certain miss cost. Comparing this model with the observed IPC gives an insight
on whether the performance is limited by the number of FPUs or by the cost of
the cache miss.

In addition to installing these dynamic probes, OMPtrace accepts static instrumentation placed by the user, for tracing of other functions or code regions
in the program. During program execution, OMPtrace generates trace records.
These records contain absolute times from the activation of the instrumented
points in the program during the parallel execution, as well as, the information
gathered for these points (for example, data from hardware performance counters). Each record represents an event or activity associated to one thread in the
system. At the end of execution, these traces are combined into a single Paraver
trace file, in order to convert these *"punctual"* events into *"interval values"*.

4   Case Study

In this section we describe the steps followed to analyze an application and
how OMPtrace and Paraver were useful to identify potential improvements. We
observe that performance tuning of any large application is in general a never-
ending task, with new potential improvements arising just after a previous one
has been implemented. Thus, our intent was not to optimize the performance
of the application to the utmost possible level. Instead, we focused on the way
Paraver was helpful in the process.

In this case study, we used the US DOE ASCI Sweep3D benchmark, which
uses a multidimensional wavefront algorithm for "discrete ordinates" determin-

istic particle transport simulation. Figure 4 displays the Sweep3D major data structures and the iteration space. The core computation presents reductions in all directions (i, j, k, and m); thus posing some problems to parallelization. To solve these problems, Sweep3D benefits from multiple wavefronts in multiple dimensions, which are partitioned and pipelined on a distributed memory system. The three dimensional space is decomposed onto a two-dimensional orthogonal mesh, where each processor is assigned one columnar domain, as shown on Figure 5(a). Sweep3D pipelines the K dimension, exchanging messages between processors as wavefronts propagate diagonally across this 3D space in eight directions.

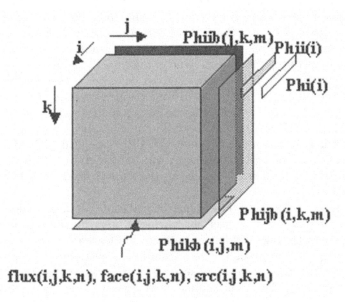

flux(i,j,k,n), face(i,j,k,n), src(i,j,k,n)

Fig. 4. Sweep3D major data structures and iteration space.

Within each MPI process domain further decomposition of work among several threads can be achieved with OpenMP. The approach is to parallelize the execution of the planes of a diagonal wavefront that traverses the sub-cube computed by each MPI process. Each such plane is inherently parallel as each of its points contributes to a different reduction in each of the i, j, and k directions, as shown in Figure 5(b). This is nevertheless at the expense of additional index computations and triangular loop trip count, which causes significant overhead both in terms of index computations and of OpenMP run time library overhead.

Figure 6 displays the computational flow of Sweep 3D in its original version, which we will refer here as "*diag*" version. However, as an alternative approach described in the source distribution, the "do idiag" and "do jkm" loops, shown

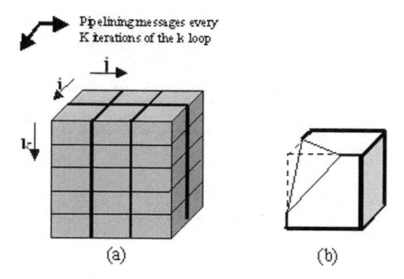

Fig. 5. (a) MPI parallelization structure and (b) sub-cube diagonal parallelization structure with OpenMP.

in Figure 6, can be replaced by a tripled nested loop ("do m", "do k", and "do j"), which we will refer here as "*mkj*" version.

Our trace collection and analysis was performed with a small problem size, using a cube of dimensions 50×50×50, running on one SP Nighthawk II node with 16 375 MHz Power3+ processors. The performance observations were then validated running a mixed MPI/OpenMP code with a larger problem size, using a cube of dimensions 300×300×100, on 12 SP Nighthawk I Nodes, each node with 8 222 MHz Power3 processors.

As described above, the original MPI version is parallelized in two levels, along the "I" and "J" dimensions. Table 1 presents the elapsed times in seconds for the MPI versions corresponding to different partitioning of the global iteration space, and the elapsed time for the OpenMP "diag" version. The first number in the decomposition indicates the number of processors used for the partitioning across the I dimension, while the second number indicates the number of processors for the J dimension.

We observed that on 6 processors, the best MPI decomposition ran in 3.69 seconds, while the OpenMP version ran in 7.78 seconds. The OpenMP performance with 12 processors was almost three times worst than the best MPI performance. In order to identify the reasons for this performance difference, we obtained two sets of trace of the MPI and the OpenMP versions, one using the default set of hardware counters to measure communication and synchronization overheads, and the other counting cache misses (level 1 and 2) and TLB misses to investigate locality issues.

```
DO iq=1,8                           ! octants
  DO mo=1,mmo                       ! angle pipelining loop
    DO kk=1,kb                      ! k-plane pipelining loop
      RECV E/W                      ! recv block I-inflows
      RECV N/S                      ! recv block J-inflows
      DO idiag=1,jt+nk-1+mmi-1      ! JK-diagonals with MMI pipelining
        DO jkm=1,ndiag              ! I-lines on this diagonal
          j,k,m = f(idiag,jkm)      ! map to j, k, and m indices
          DO i=1,it                 ! source (from Pn moments)
          ENDDO
          DO i=i0,i1,i2             ! Sn eqn
          ENDDO
          DO i=1,it                 ! flux (Pn moments)
          ENDDO
          DO i=1,it                 ! DSA face currents
          ENDDO
        ENDDO
      ENDDO
      SEND E/W                      ! send block I-outflows
      SEND N/S                      ! send block J-outflows
    ENDDO
  ENDDO
ENDDO
```

Fig. 6. Sweep 3D control flow

Table 1. Elapsed time in seconds for "diag", running the small problem with OpenMP and with MPI.

NB Domains	OpenMP time	Decomposition	MPI time
6	7.78	1x6	3.97
		2x3	3.69
		3x2	3.71
		6x1	4.47
12	6.55	1x12	3.50
		2x6	2.74
		3x4	2.21
		4x3	2.25
		6x2	2.97
		12x1	3.98

Using Paraver to compute the total useful computation, we observed that both versions were loosing a similar percentage of time in synchronization and communication. The percentage of time inside numerical computation routines was around 65% for both runs. In the case of OpenMP this low percentage was partially due to the fine granularity of the triangular loops, and because it still executed some sequential computation, since only the major computational loop was parallelized. On the other hand, we observed that the OpenMP version had less L1 misses (2137 per ms) and TLB misses than the MPI version (4153 L1 misses per ms), but much more L2 misses (1356 per ms for the OpenMP version versus only 133 per ms for the MPI version). The rate of L2 misses per millisecond for one traversal of the 3D iteration space in "diag" is shown in Figure 7 and Figure 8 for the OpenMP and MPI versions respectively. In these figures, darker gray (blue) represents large values, while white (green) corresponds to low values. The areas with low values in Figure 7 correspond to intervals where the threads are waiting for work or synchronizing. Hence, our optimization efforts concentrated on improving locality of the OpenMP version and minimizing coherence invalidations.

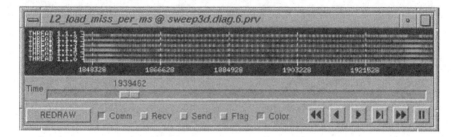

Fig. 7. L2 misses per milliseconds for the OpenMP execution of "diag".

Fig. 8. L2 misses per milliseconds for the MPI execution of "diag".

Taking into account that shared memory is inherently more efficient than message passing for fine grain synchronization, we implemented an OpenMP parallel version based on the "mkj" version, where the outer loop was parallelized and the internal precedence was enforced by some synchronization mechanism. Two approaches were implemented. The first was the version "*ccrit*", which uses the CRITICAL OpenMP directive for the implementation of the reduction. The result was a fairly high contention on the lock, a behavior that could be visualized with Paraver, as shown in Figure 9, which displays the behavior of the critical section access. The long regions in gray (red) correspond to the time threads are trying to get a lock that is already taken. The dark (blue) periods correspond to threads using the lock. White (green) is when a thread releases the lock and light gray (light blue) corresponds to execution outside of the critical section. As can be observed, the sequence of accesses does not follow a specific pattern, and the waiting time to obtain a lock has a large variance. Using the quantitative analysis module of Paraver we measured the average waiting time to be 170 microseconds with a standard deviation of 172.

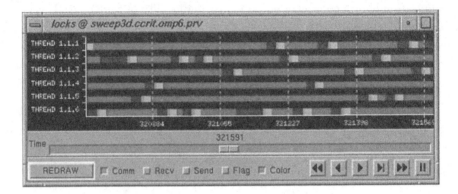

Fig. 9. Lock access pattern in the "ccrit" version.

The second synchronization mechanism, which we refer to "*cpipe*", was implemented based on shared arrays and busy waiting. In this version the iterations of the parallelized loop ("do m") are interleaved across threads. When a thread finalizes its part of computation involved in the reduction it signals to the following thread to continue with its part. In this approach, the reductions are executed in sequential order (although different threads compute different parts). Thus, in this case, since the sequential order of the computation of the reductions are preserved, the numerical results are identical to the sequential execution, independently of the number of processors used in the parallel computation. This is an important advantage compared to the "ccrit" version, where the critical sections used to assure atomicity of the reduction updates do not preserved their order; thus resulting in numerical differences between runs. In this version we observed that the synchronization overhead was very low and a fairly good pipelining was achieved.

A comparison of the single processor run of the "diag" and "mkj" versions showed that the "mkj" version was significantly slower. Using Paraver traces with the sequential application we observed that the first, third and fourth "do i" loops were touching the variables "flux", "src", and "face" and incurred most of the level 2 cache misses. However, analyzing the source code, it could be observed that interchanging the "do m" loop inwards would reduce misses in the "do i" loops. Besides the locality problems, parallelizing the m dimension also has the problem of the small trip count of the loop (only 6), which limits parallelism. Taking into account data locality and trip count considerations described above, we interchanged the loops, creating the version "*kjmi*". In order to achieve good pipelining overlap in this version, the "do k" loop was parallelized with SCHEDULE(STATIC,1), which means totally interleaved. This causes matrix "Phikb" to be circulated between processors for each k, generating level 2 cache invalidations and a slightly higher miss ratio than the MPI version. Thus, in order to increase the reuse of "Phikb" we introduced a final modification (version "*Kjkmi*") where the k loop was strip-mined and interchanged according to the version name.

Table 2. Elapsed time in seconds for the different OpenMP versions.

version	1	2	3	4	5	6	8	9	10	11	12	13	14	15	16
ccrit	28.26	24.41	26.84	26.47	29.28	30.34					30.43				
cpipe	25.63	18.45	13.01	12.53	10.06	7.67					7.76				
diag	17.28	13.09	11.40	9.64	8.50	7.78					6.55				
kjmi	14.86	10.01	7.35	5.82	4.89	4.34	3.62	3.38	3.09	3.04	2.88	2.69	2.69	2.64	2.53
Kjkmi	14.91	8.47	6.35	4.91	4.24	3.58	2.90	2.81	2.78	2.65	2.29	2.22	2.16	2.19	2.15

A performance summary of the different OpenMP versions of the program, running the small problem size is presented in Table 2. The numbers show the inefficiency and lack of scalability of the "ccrit" version. The overhead of the mutex lock and unlock needed to protect the critical section can be observed, when comparing the times for just one thread in the "ccrit" and "cpipe" versions. The huge contention at the lock shown in Figure 9 causes the scalability problems.

Although "cpipe" performed better than "ccrit", when comparing to the other three versions we observe that "cpipe" also had poor locality behavior, and scalability problems. These problems occur mainly because the parallel loop on "m" has only an iteration count of 6, which results in a poor pipelining.

The scalability of "diag" is limited, as mentioned above, due to the very high number of L2 misses caused by false sharing and the variable trip count of the parallelized loops. The overhead of opening and closing such parallel loops with very small trip counts for the diagonal planes at the corners of the cube also contributes to the poor scalability of this version.

The two final versions show much better behavior for just one thread, an effect that not only benefits the OpenMP code, but also the MPI. Scalability is fair, and the performance achieved is equivalent to that of pure MPI as reported in Table 1. In some cases, as "Kjkmi" running 6 threads, as well as in other experiments we have performed with larger problem sizes, we observed that the OpenMP versions were marginally better than the pure MPI version.

Table 3 presents the best elapsed time over two runs using the larger problem size (300×300×100) on 12 IBM SP Nighthawk I Nodes at Lawrence Livermore National Laboratory. Notice that this table shows only a few combinations of MPI tasks and OpenMP threads that were selected from the large space of

Table 3. Elapsed time in seconds, for the large problem size, running on 12 IBM SP Nighthawk I Nodes.

MPI Tasks	Version	Threads							
		0	2	3	4	5	6	7	8
12	kjmi					56.61	55.33	52.96	40.55
	Kjkmi					56.62	57.00	56.70	57.14
24	kjmi			49.87	40.39				
	Kjkmi			65.69	45.61				
48	kjmi		39.41						
	Kjkmi		45.69						
84	diag	41.68							
96	diag	40.93							

possible configurations. Also, for each MPI task, only one decomposition was considered. Therefore, based on the results from the small problem size, where we observed that the MPI performance is heavily dependent on the decomposition, one should be aware that the MPI decomposition chosen for these experiments, was based on the observations from the small problem size, but the times might not necessarily represent the best MPI performance for this problem size.

We observe that when running the large problem size with all 96 processors, we were able to confirm the analysis derived from the small problem size for the mixed MPI/OpenMP version of "kjmi". This version performed slightly better than the pure MPI version with all three combinations of tasks and threads (namely, 48/2, 24/4, and 12/8). On the other hand, the performance of the "Kjkmi" version did not perform as well as expected, and in the only situation where its performance was comparable to the "kjmi" version, it did dot scale well with more than 5 threads. Thus, more analysis is necessary to understand its performance behavior.

When running the mixed mode approach, we observed some conflicts in the scheduling of the two parallelization strategies (MPI and OpenMP). This problem can be observed in Figure 10, which shows for each thread, the pattern of computation from the iterations of the parallel loop, when running "kjmi" with the small problem size, using 2 MPI tasks and 8 OpenMP threads per task. In this figure, where each dark (blue) area between two flags corresponds to one iteration of the loop, we can observe an unbalance between threads inside of each MPI task. The reason for this unbalance is due to the MPI pipelining that was set to have "k" dimension of 10 planes. Hence, the parallel loop had only 10 iterations, and when scheduling such number of iterations among 8 threads, two of then will perform two iterations, while the other six threads will perform only one iteration and then wait for the first two to finish.

Therefore, another important observation of this experiment was that when mixing different programming models, it is of key importance to analyze the scheduling decisions taken by the different parallelization strategies. It is fairly easy for these strategies to interfere with each other, and without an analysis

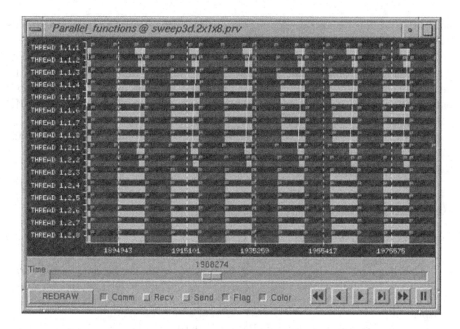

Fig. 10. Scheduling of the OpenMP loop iterations

tool such as OMPtrace and Paraver it may be difficult to understand the effects in performance.

5   Conclusions

In this paper we described OMPtrace, a dynamic tracing mechanism that combines traditional tracing with dynamic instrumentation and access to hardware performance counters to create a powerful tool for performance analysis and optimization of OpenMP applications. Performance data collected with OMPtrace is used as input to the Paraver visualization tool for detailed analysis of the parallel behavior of applications.

The usefulness of OMPtrace and the power of Paraver for tuning and optimizing OpenMP applications was illustrated in a case study with the US DOE ASCI Sweep3D benchmark. We analyzed the performance of a small problem size of Sweep3D, running on a single IBM SP node with 16 processors, and validated the performance observations and code optimizations, running a mixed MPI/OpenMP version of the code, with a larger problem size, on 12 IBM SP Nodes with 8 processors each.

The performance of the original OpenMP version was three times slower than the MPI version when running on 12 processors of a single IBM SP node, but when running the optimized version with the larger problem size on 96 processors, the mixed MPI/OpenMP version performed slightly better than the pure MPI version for all three combinations of MPI tasks and OpenMP threads used (48 task and 2 threads, 24 tasks and 4 threads, and 12 tasks and 8 threads).

We notice that the two dimensional MPI parallelization and the corresponding combinations of possible decompositions had an important effect on the performance of the MPI program. In OpenMP there is only one dimensional parallelization and it would be interesting to experiment with nested OpenMP parallelism.

Finally, we observe that there are still several issues, such as memory management and scheduling conflicts, that remain as challenges for optimization of the mixed MPI/OpenMP application.

References

1. B. R. Buck and J. K. Hollingsworth. An API for Runtime Code Patching. In *Journal of High Performance Computing Applications*, 14(4):317–329, Winter 2000.
2. L. DeRose and T. H. Hoover Jr. and J. K. Hollingsworth The Dynamic Probe Class Library - An Infrastructure for Developing Instrumentation for Performance Tools. In *Proceedings of 2001 International Parallel and Distributed Processing Symposium*, April 2001.
3. European Center for Parallelism of Barcelona (CEPBA). *Paraver - Parallel Program Visualization and Analysis Tool - Reference Manual*, November 2000. http://www.cepba.upc.es/paraver.
4. K. R. Koch, R. S. Baker, R. E. Alcouffe. Solution of the First-Order Form of the 3-D Discrete Ordinates Equation on a Massively Parallel Processor. In *Trans. Amer. Nuc. Soc.* 65(198), 1992.
5. B. P. Miller, M. D. Callaghan, J. M. Cargille, J. K. Hollingsworth, R. B. Irvin, K. L. Karavanic, K. Kunchithapadam, and T. Newhall. The Paradyn Parallel Performance Measurement Tools. In *IEEE Computer*, 28(11):37–46, November 1995.

A Comparison of Scalable Labeling Schemes for Detecting Races in OpenMP Programs*

So-Hee Park, Mi-Young Park, and Yong-Kee Jun**

Dept. of Computer Science, Gyeongsang National University
Chinju, 660-701 South Korea
{shpark,park,jun}@race.gsnu.ac.kr

Abstract. Detecting races is important for debugging shared-memory parallel programs, because the races result in unintended nondeterministic executions of the program. On-the-fly technique to detect races uses a scalable labeling scheme which generates concurrency information of parallel threads without any globally-shared data structure. Two efficient schemes of scalable labeling, *BD Labeling* and *NR Labeling*, show the similar complexities in space and time, but their actual efficiencies have been compared empirically in no literature to the best of our knowledge. In this paper, we empirically compare these two labeling schemes by monitoring a set of OpenMP kernel programs with nested parallelism. The empirical results show that *NR Labeling is more efficient than BD Labeling* by at least 1.5 times in generating the thread labels, and by at least 3.5 times in using the labels to detect races in the kernel programs.

1 Introduction

A *race* is a pair of instructions which accesses a shared variable with at least one write access without coordination in an execution of the program. Detecting the races is important for debugging such a kind of parallel programs as OpenMP programs [2,7], because the races result in unintended non-deterministic executions of the program. *On-the-fly race detection* [1,3,4,6] instruments a debugged program and monitors an execution instance of the program to report races which occur during the monitored execution. To detect any race which involves the current access in an execution, this technique uses a *race detection protocol* which determines the logical concurrency between the current access and the previous conflicting accesses, and then maintains an *access history* of the shared variable for the subsequent race detection. For the concurrency information to be used in the protocol, an on-line *labeling scheme* generates the logical concurrency information, called a *label*, for every thread in the execution.

Various on-line labeling schemes [1,3,4,6] have been reported in the literatures, but some labeling schemes [3] use a centralized data structure which is

---

* This work is supported by University Research Program supported by Ministry of Information and Communication in South Korea.
** In Gyeongsang National University, he is also involved in both the Institute of Computer Research and Development, and the Information and Communication Research Center, as a research professor.

globally-shared among the threads and then may incur serious bottleneck in generating the thread labels. Among the scalable schemes, *OS Labeling* [6] is less efficient than the other scalable schemes [1,4], because it increases the storage space of each label in the access histories in proportion to the nesting depth of the monitored parallelism and then requires an additional time overhead of garbage collection to save the space for the access histories. The other two scalable labeling schemes, *BD Labeling* [1] and *NR Labeling* [4], generate a constant-sized label for each entry of access histories, and show the similar complexities each other in space and time, but their actual efficiencies have been compared empirically in no literature to the best of our knowledge.

This paper compares the actual efficiencies of these two scalable schemes: BD Labeling and NR Labeling. BD Labeling generates a label on every event to fork or join or coordinate threads, because its basic idea was originated from monitoring distributed-memory programs with nested parallelism. To monitor an execution of OpenMP program, it is sufficient for each logical thread to have just one label, such as in NR Labeling. Thus we modify BD Labeling to be more efficient by making it generate only one label for each thread, and call it *Thread-based-BD (T-BD) Labeling*. We empirically compare T-BD Labeling with NR Labeling using a set of OpenMP kernel programs with nested parallelism. The kernel benchmark programs are monitored sequentially, because the comparison requires to measure the net execution time of labeling scheme in a monitored execution of program. The empirical results show that NR Labeling is still more efficient than T-BD Labeling in every aspect of time efficiencies in on-the-fly race detection.

After describing first the background information on this work in section 2, we introduce the two scalable labeling schemes, T-BD Labeling and NR Labeling, with the corresponding examples in the following two sections. Then we explain our experimentation which includes the race detection protocol we used, the kernel programs, and the measurement strategy before analyzing the results. Our work is concluded in the final section.

2   Background

This section describes the OpenMP nested parallelism considered in this work using a directed acyclic graph called *Partial Order Execution Graph* (POEG) [3]. POEG captures the *happened-before* [5] relationship and imposes a partial order on the set of threads that make up an execution instance. Then we discuss the labeling schemes to generate information of such concurrency relationship to detect races in an execution of parallel programs.

2.1   OpenMP Program

In this paper, we consider OpenMP programs with nested parallel loops without coordination operation between threads. If there is no other loop contained in a loop body, the loop is called an *innermost* loop. Otherwise, it is called an *outer* loop. In a nested do-loop D, an individual loop can be enclosed by many outer

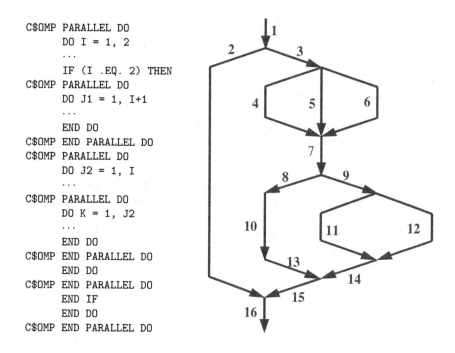

```
C$OMP PARALLEL DO
      DO I = 1, 2
        ...
      IF (I .EQ. 2) THEN
C$OMP PARALLEL DO
      DO J1 = 1, I+1
        ...
      END DO
C$OMP END PARALLEL DO
C$OMP PARALLEL DO
      DO J2 = 1, I
        ...
C$OMP PARALLEL DO
      DO K = 1, J2
        ...
      END DO
C$OMP END PARALLEL DO
      END DO
C$OMP END PARALLEL DO
      END IF
      END DO
C$OMP END PARALLEL DO
```

Fig. 1. An OpenMP Parallel Program and its POEG

loops. The *nesting level* of an individual loop is equal to one plus the number of the enclosing outer loops. The *nesting depth* of D is the maximum nesting level of loops in D. In a *perfectly* or *one-way* nested loop of nesting depth N, there is exactly one loop at each nesting level i, $(i = 1, 2, \cdots, N)$. A loop is *multi-way* or *m-way* nested if there exist m disjoint loops in a nesting level, $(m \geq 1)$. For example, Figure 1 shows a two-way nested loop of nesting depth three. If we remove the loop indexed by J1 in the figure, the program becomes a one-way nested loop of nesting depth three. Let I_i denote the loop index of a loop D_i, L_i and U_i denote the lower and upper bound of I_i respectively, and A_i denote the increment of I_i. A loop D_i is *normalized* if the values of both L_i and A_i are one. A_i is optional when it is equal to one. To make our presentation simple, we assume that all parallel loops are normalized loops. Figure 1 shows a normalized loop with index J1 for which the lower bound is one, the upper bound is I+1, and the increment is one.

In an execution of OpenMP program, more than one thread can be *forked* to share a work at a 'C$OMP PARALLEL DO' directive, and can be *joined* at the corresponding 'C$OMP END PARALLEL DO' directive. Such fork and join operations are called *thread operations*. Figure 1 shows an OpenMP Fortran program where four parallel loops are specified with the corresponding pairs of directives. The concurrency relationship among threads is represented by a directed acyclic graph, called *Partial Order Execution Graph (POEG)* [3]. A vertex of POEG represents a thread operation, and an arc started from a vertex represents a *logical thread* started from the corresponding thread operation. Since the graph captures the *happened-before* relationship [5], it presents a partial order on a set of the logical threads that make up an execution instance of a parallel program. For example, the POEG in Figure 1 shows a partial order on the threads in an execution instance of the two-way nested loop.

Ordering threads does not depend on the number and the relative execution speeds of processors that executes the program. A thread t_a *happened before* a thread t_b, if there exists a path from t_a to t_b in the graph. In this case, t_a is an *ancestor thread* of t_b, and t_b is a *descendant thread* of t_a. A thread t_a (t_b) is a *parent* (*child*) thread of t_b (t_a), if there exists no thread t_x such that t_a (t_b) happened before t_x which happened before t_b (t_a). There exists two *concurrent threads* if and only if there exists no path between the two threads in the POEG. For example, consider the execution instance of POEG shown in Figure 1. In this figure, the thread numbered three, denoted by t_3, happened before the thread t_9, because there exist a path from t_3 to t_9. The thread t_2 is concurrent with t_8, because there exist no path between the two threads.

2.2   Online Labeling Schemes

On-the-fly race detection instruments a debugged program and monitors an execution instance of the program to report races which occur during the monitored execution. To detect any race which involves the current access in an execution, this technique uses a *race detection protocol* which determines the logical concurrency between the current access and the previous conflicting accesses, and then maintains an *access history* of the shared variable for the subsequent race detection. For the concurrency information to be used in the protocol, an on-line *labeling scheme* generates the logical concurrency information, called a *label*, for every thread in the execution.

Various on-line labeling schemes [1,3,4,6] have been reported in the literatures, but some labeling schemes such as *Task Recycling* [3] use a centralized data structure which is globally-shared among the threads and then may incur serious bottleneck in generating the thread labels. Among the scalable schemes, *OS Labeling* [6] is less efficient than the other scalable schemes [1,4], because it increases the storage space of each label in the access histories in proportion to the nesting depth of the monitored parallelism and then requires an additional time overhead of garbage collection to save the space for the access histories. The other two scalable labeling schemes, *BD Labeling* [1] and *NR Labeling* [4], generate a constant-sized label for each entry of access histories, and show the similar complexities each other in space and time, but their actual efficiencies have been compared empirically in no literature to the best of our knowledge.

This paper compares the actual efficiencies of these two scalable schemes: BD Labeling and NR Labeling. BD Labeling generates a label on every event to fork or coordinate threads in addition to every logical thread, because its basic idea was originated from monitoring distributed-memory programs with nested parallelism. To monitor an execution of shared-memory OpenMP program, it is sufficient for each logical thread in POEG to have just one label, such as in NR

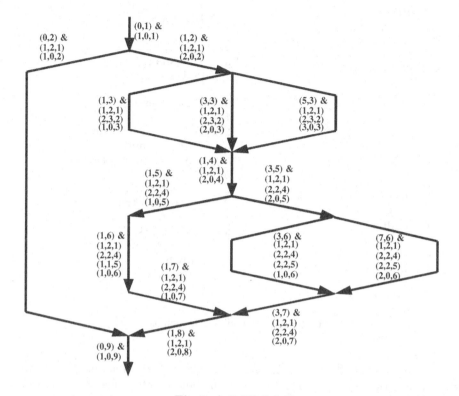

Fig. 2. A T-BD Labeling

Labeling. For example, we generate only one label for a logical thread which is started from a join operation and terminated at a fork operation. BD Labeling however generates two labels for this thread: one is generated for the thread itself immediately after the join operation is completed and the other is generated for the fork operation. Thus we modify BD Labeling to be more efficient by making it generate only one label for each thread, and call it *Thread-based-BD* (*T-BD*) Labeling. This thread-based labeling scheme can also be applied to the coordination operations between threads in a similar way.

3   T-BD Labeling Scheme

In T-BD Labeling, each logical thread has concurrency information which consists of two main components: a *BD label* and a list structure L, denoted by BD & L altogether. A BD label identifies a thread t_n in the nesting level n with two subcomponents: $BD_n = (b_n, d_n)$, where b and d are integers called a *breadth* and a *depth* of the thread, respectively. The list L_n of t_n is a list of triples. The i-th entry of L_n, denoted $L_n(i)$, consists of three integers, (t_{i-1}, T_i, d_{i-1}), for the most recent thread t_{i-1} in nesting level $i - 1$ such that t_{i-1} is ordered

```
0  T-BD-Init()
1      i := 0;
2      (b, d) := (0, 1);
3      L(1) := (1, 0, 1);
4  End T-BD-Init
```

```
0  T-BD-Fork()
1      i := i_p + 1;
2      for a := 1 to i do
3          L(a) := L_p(a);
4      endfor
5      L(i, T) := U;
6      g := t - 1;
7      for k := 1 to i - 1 do
8          g := g × L(k, T);
9      endfor
10     b := b_p + g;
11     d := d_p + 1;
12     L(i + 1) := (t, 0, d);
13     d' := d;
14 End T-BD-Fork
```

```
0  T-BD-Join()
1      d := d' + 1;
2      L(i + 1, d) := d;
3      d'_p := max{d'_p, d};
4  End T-BD-Join
```

```
0  T-BD-Ordered(b, d)
1      k := 1;
2      while L(k, T) ≠ 0 do
3          t_b := b mod L(k, T) + 1;
4          if t_b ≠ L(k + 1, t) then
5              return false;
6          elseif d ≤ L(k + 1, d) then
7              return true;
8          endif
9          b := b/L(k, T);
10         k := k + 1;
11     endwhile
12     return true;
13 End T-BD-Ordered
```

Fig. 3. T-BD Labeling Algorithms

with the thread t_n, where $i = 1, 2, \cdots, n + 1$. Here, t_i is the loop index of thread which is less than or equal to T_i which is the number of siblings of the thread t_i. And the thread depth d_i is the *Lamport timestamp* [5] which means one plus the number of threads that happened before the thread t_i in the critical path from the initial thread in the POEG. The breadth b_i of a thread t_i is defined as $b_i = b_{i-1} + (t_i - 1) \prod_{k=1}^{i-1} L_i(k, T)$, where $b_0 = 0$ and $L_i(k, T)$ stands for T_i in $L_i(k)$. From this equation, we have $b_1 = t_1 - 1$, which is the remainder of dividing b_i with $L_i(1, T)$.

For example, consider a thread t_3 in Figure 2 whose BD label is $(3, 6)$ in the third nesting level. The breadth of this thread b_3 is three, because $b_0 = 0$, $b_1 = (b_0 + 1) = 1$, $b_2 = (b_1 + 1 \times 2) = 3$, and $b_3 = (b_2 + 0) = 3$. $L_3(4, T)$ of this thread is zero, because the thread does not know the number of threads to be forked next. This entry is only for initializing t_3 and d_3 in $L_3(4)$. Whenever a thread in the nesting level i is forked, the on-line algorithm computes $\{b_i, d_i, T_i\}$ to generate its BD label, $L_i(i, T)$, and $L_i(i+1)$ but $L_i(i+1, T)$. In this process, a constant number of computations is sufficient except the case of computing b_i whose number of computations increase as large as the size of L_i which increases as large as the nesting depth.

When we check concurrency relationship between any two threads, t_i and t_j, only one of their L_i and L_j is sufficient to do the job. Consider the case of using L_j. By the definition of the breadth of a thread, a thread t_m can be assumed as t_i, where $m = 1, 2, \cdots, i$, from obtaining one plus the remainder in dividing b_i by $L_j(k, T)$, where $k = 1, 2, \cdots, j$. We compare t_m with t_k in $L_j(k + 1, t)$. If t_m is equal to t_k, and d_i is less than or equal to d_k in $L_j(k + 1, d)$, the thread t_i

happened before t_j. For example, consider the two threads in Figure 2: $t_i = (5,3)$ and $t_j = (3,6)$. In case we use the list structure L_3 of t_j, we can compute t_m using $b_i = 5$ and $L_j(k, T)$, where $k = 1, \cdots, 3$, resulting in $t_m = 2$, where $m = 1$. Therefore, the thread t_i happened before t_j, because the computed $t_1 = 2$ with $b_i = 5$ is equal to the stored $t_1 = 2$ in $L_3(2, t)$, and $d_i = 3$ is less than $d_1 = 4$ in $L_3(2, d)$.

We implemented T-BD Labeling with four algorithms shown in Figure 3: T-BD-Init(), T-BD-Fork(), T-BD-Join(), and T-BD-Ordered(). Among them, T-BD-Ordered(b, d) checks the logical concurrency between the current thread and another thread whose label is (b, d). The other three algorithms generate concurrency information including BD label: T-BD-Init() for an initial thread, T-BD-Fork() for a thread started from fork operation, and T-BD-Join() for a thread started from join operation. In the algorithms, each data structure with a subscript p such as L_p means the corresponding data structure of the ancestor thread which forked the current thread. The depth variable d' is a mirror variable which is shared locally by the siblings of the current thread to help maintain d in the thread. And U denotes the upper bound of the current loop index.

4   NR Labeling Scheme

In NR Labeling, each thread has concurrency information which consists of three main components: the *one-way depth* ξ, the *one-way region* OR, and the *one-way history* OH of the thread, denoted by $[\xi, OR] \bowtie OH$ altogether. The *one-way depth* ξ is one plus the nesting level of the most recent *one-way root* of the thread. A *one-way root* of a thread t is the most recently joined ancestor of t in each nesting level which is higher than or equal to the nesting level of t. Here the set of joined ancestors of a thread are assumed to include the initial thread and the thread itself if it is a joined thread. The *one-way region* OR of a thread is a pair of a *join counter* λ and a *nest region* $\langle \alpha, \beta \rangle$, denoted by $[\lambda, \langle \alpha, \beta \rangle]$ altogether. The join counter λ of a thread is the number of the joined ancestors of the thread in the critical path from the initial thread. The nest region consists of two integers $\langle \alpha, \beta \rangle$ which mean a range of number space divided by each fork operation and concatenated by each join operation. This one-way region of a thread can be used as the thread identifier, and stored in access histories as an NR label of the thread for detecting races. The *one-way history* OH of a thread is an ordered list of the one-way roots of the thread which are represesed by their one-way regions in an ascending order of their join counters.

Figure 4 shows an example of NR Labeling. The nest region of the initial thread is $\langle 1, 50 \rangle$ just for readability, although it is initialized with $\langle 1, maxint \rangle$ in general where $maxint$ is the maximum integer that can be represented in a machine. This initial nest region is divided in two regions in the child threads of the initial thread into $\langle 1, 25 \rangle$ and $\langle 26, 50 \rangle$, and concatenated back to the initial region in the last thread. For an example of labeled thread, consider a thread t in the lower part of the figure whose one-way region is $OR_t = [4, \langle 26, 50 \rangle]$. The join counter of t, λ_t, is four, because t has four joined ancestors in the critical path from the initial thread: the threads whose one-way regions are $[1, \langle 1, 50 \rangle]$

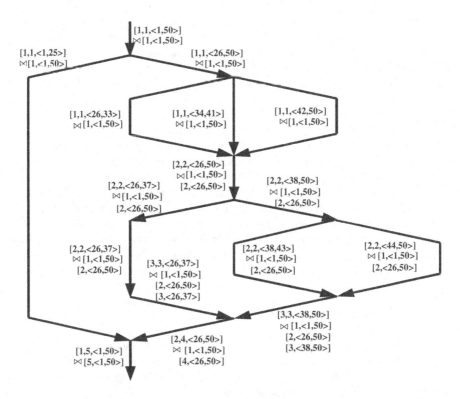

Fig. 4. An NR Labeling

of the initial thread, $[2, \langle 26, 50 \rangle]$, $[3, \langle 26, 37 \rangle]$ or $[3, \langle 38, 50 \rangle]$, and $[4, \langle 26, 50 \rangle]$ of t. Therefore the one-way history of t, OH_t, is the ordered list of two one-way roots of t: $[1, \langle 1, 50 \rangle]$ in the nesting level of zero, and $[4, \langle 26, 50 \rangle]$ of t in the nesting level of one. And the one-way depth of t, ξ_t, is two, because t itself is the most recent one-way root of t and the nesting level of t is one. For the details of NR Labeling, refer to the original paper [4].

To compare the concurrency relationship between any two threads, t_i and t_j, we need OR_i, OR_j, and only one of OH_i and OH_j. Here we consider the case of using OH_j. For this case, we first have to figure out the *nearest one-way root* of t_j to t_i, which is a thread t_x in OH_j such that λ_x is the smallest join counter in OH_j that is greater than λ_i. Then, the thread t_i happened before t_j, if it satisfies $\lambda_i < \lambda_x \leq \lambda_j$ and the nest region $\langle \alpha_i, \beta_i \rangle$ overlaps with $\langle \alpha_x, \beta_x \rangle$ in OH_j, or it satisfies $\lambda_i = \lambda_j$ and $\langle \alpha_i, \beta_i \rangle$ overlaps with $\langle \alpha_j, \beta_j \rangle$. Here, we can use a binary search to find out t_x in OH_j, because an OH is a list ordered with the join counter as the search key. For example, consider the two threads, t_a and t_b in the Figure 4, whose labels are $OR_a = [1, \langle 26, 33 \rangle]$ and $OR_b = [3, \langle 38, 50 \rangle]$ with its OH_b. In this case, the nearest one-way root of t_b to t_a is the thread t_c whose one-way region $OR_c = [2, \langle 26, 50 \rangle]$. Therefore, t_a and t_b are ordered each other,

```
0  NR-Init()                          0  NR-Join()
1     ⟨α, β⟩ := ⟨1, maxint⟩;          1     λ := λ' + 1;
2     λ := 1;                          2     if (OH(ξ, ⟨α, β⟩)) = ⟨α, β⟩) then
3     OH(1) := [λ, ⟨α, β⟩];           3        OH(ξ, λ) := λ;
4     ξ := 1;                          4     else ξ := ξ + 1;
5  End NR-Init                         5        OH(ξ) := [λ, ⟨α, β⟩];
                                       6     endif
0  NR-Fork()                           7     λ'_p := max{λ'_p, λ};
1     stride := (β_p − α_p + 1)/U;    8  End NR-Join
2     α := α_p + (I − 1) × stride;
3     if (i < U) then                 0  NR-Ordered(OR_i)
4        β := α + stride − 1;         1     if (λ_i < λ) then
5     else β := β_p;                   2        OR := NR_BinarySearch(OH, λ_i, ξ);
6     endif                           3        if (α_i ≤ OR(β) and OR(α) ≤ β_i) then
7     λ := λ_p;                        4           return true;
8     for a := 1 to ξ do              5        endif
9        OH(a) := OH_p(a);            6     elseif (α_i ≤ β and α ≤ β_i) then
10    endfor                          7        return true;
11    ξ := ξ_p;                       8     endif
12    λ' := λ;                        9     return false;
13 End NR-Fork                        10 End NR-Ordered
```

Fig. 5. NR Labeling Algorithms

because the three join counters satisfy $(1 < 2 < 3)$ and the nest region $\langle 26, 33 \rangle$ overlaps with $\langle 26, 50 \rangle$.

We implemented NR Labeling with four algorithms shown in Figure 5: NR-Init(), NR-Fork(), NR-Join(), and NR-Ordered(). NR-Ordered(OR_i) checks the logical concurrency between the current thread and another thread whose label is OR_i. The other three algorithms generate concurrency information including NR label: NR-Init() for an initial thread, NR-Fork() for a thread started from fork operation, and NR-Join() for a thread started from join operation. In the algorithms, each data structure with a subscript p such as OH_p means the corresponding data structure of the ancestor thread which forked the current thread. The counter variable λ' is a mirror variable which is shared locally by the siblings of the current thread to help maintain λ in the thread. And U denotes the upper bound of the current loop index I.

5   Empirical Comparison

In this section, we empirically compare T-BD Labeling with NR Labeling using a set of OpenMP kernel programs with nested parallelism. For race detection protocol that uses the two labeling schemes, we chose Mellor-Crummey's protocol [6] which is efficient to monitor a parallel program with nested parallelism and no other inter-thread coordination. This protocol is efficient, because it requires only three entries in an access history of a shared variable. Figure 6 shows the Mellor-Crummey's protocol implemented for this work, where we use $AH(X)$ to denote

```
0 CheckRead(X, Label)
1    if ¬Ordered(AH(X, W)) then
2        report a race;
3    endif
4    if AH(X, R_L) = ∅ or
        LeftOf(AH(X, R_L)) or
        Ordered(AH(X, R_L)) then
5        AH(X, R_L) := Label;
6    endif
7    if AH(X, R_R) = ∅ or
        ¬LeftOf(AH(X, R_R)) or
        Ordered(AH(X, R_R)) then
8        AH(X, R_R) := Label;
9    endif
10 End CheckRead
```

```
0 CheckWrite(X, Label)
1    if ¬Ordered(AH(X, W)) then
2        report a race;
3    endif
4    if ¬Ordered(AH(X, R_L)) or
        ¬Ordered(AH(X, R_R)) then
5        report a race;
6    endif
7    AH(X, W) := Label;
8 End CheckWrite
```

Fig. 6. The Mellor-Crummey Protocol

the access history of shared variable X. $AH(X)$ consists of three entries: one entry for write access, $AH(X, W)$, and two entries for concurrent read accesses, $AH(X, R_L)$ and $AH(X, R_R)$. $AH(X, R_L)$ and $AH(X, R_R)$ denote the leftmost and rightmost thread respectively that are distinguished using the *left-of* relation [6]. The left-of relation establishes a canonical ordering of two relative threads with respect to their thread indexes of a loop from which ancestors of the two threads are forked. This protocol therefore especially requires such a labeling scheme that can evaluate the left-of relation for any two threads. Figure 7 shows two algorithms to evaluate the left-of relation for the two labeling schemes.

The efficiency of on-the-fly race detection is classified into space and time complexity. Space complexity consists of the space to store access histories for all shared variables and the space to store thread labels of simultaneously active threads. Time complexity consists of the time to generate a label for every thread and to determine a race and maintain the access history in every access to a shared variable. Here, the time to determine a race includes the time to determine if the two threads of the conflicting accesses are ordered. Using the Mellor-Crummey protocol, both T-BD Labeling and NR Labeling show the same worst-case complexity of $O(V + NT)$ in space to store constant-sized labels in all access histories and to store all the concurrency information of threads each of which size is $O(N)$, where V is the number of monitored shared variables, and N and T are the nesting depth and the maximum parallelism of the monitored program, respectively. Both T-BD Labeling and NR Labeling create each thread label in time of $O(N)$ in the worst-case, but they are different in time to determine the logical concurrency between any two threads, which are $O(N)$ and $O(\log_2 N)$ for T-BD Labeling and NR Labeling respectively. They are also different in the worst-case complexities of time to determine the left-of relation between any two threads, which are $O(N)$ and $O(1)$ for T-BD Labeling and NR Labeling respectively. This amount of difference in time, however, can be trivial, because N is typically as very small as nine or ten even in some serious large-scale

```
0  T-BD-LeftOf(b, d)
1     k := 1;
2     while L(k, T) ≠ 0 do
3         t_b := b mod L(k, T) + 1;
4         if t_b > L(k + 1, t) then        0  NR-LefOf(OR_i)
5             return true;                 1     if β < α_i then
6         endif                            2         return true;
7         b := b/L(k, T);                  3     endif
8         k := k + 1;                      4     return false;
9     endwhile                            5  End NR-LeftOf
10    return false;
11 End T-BD-LeftOf
```

Fig. 7. The Left-Of Algorithms of T-BD and NR Labeling

applications. We use Mellor-Crummey's detection protocol because, as shown in Figure 6, the protocol requires a constant time to detect races in each access except these two kinds of efficiencies from labeling scheme: the times to determine the logical concurrency and the left-of relation between any pair of threads.

As of writing this paper, we do not have any published OpenMP benchmark program that is appropriate to evaluate labeling schemes for detecting races in OpenMP programs. Omni-OpenMP team [8] developed some OpenMP versions of NAS benchmark programs, but they are neither appropriate to evaluate on-the-fly race detection tools nor have nested parallelism considered in this work. To measure the actual efficiencies of T-BD Labeling and NR Labeling, we use a set of OpenMP kernel programs which are compiled by Omni OpenMP Compiler Version 1.2.0 [8], and run on Intel Pentium-III PC under Linux Red Hat Version 6.2. We have two sets of OpenMP-Fortran kernel programs: one set of programs with no nested loops but one-way nested loops and the other set of programs with no nested loops but two-way nested loops. Each kernel program executes ten accesses to one shared variable in every thread. The characteristics of our kernel programs is shown in the corresponding columns of Table 1 and 2: the nesting depth $N = \{3, 4, 5\}$, the number of threads forked in every parallel loop $t = \{10, 15, 20\}$, the names of programs, and the original execution time that is measured in unmonitored phase.

We instrumented the kernel programs for the two kinds of algorithms: the labeling schemes to generate concurrency information on every thread in an execution of the program, and the race detection protocol on every access to detect races occurred in the execution instance. These algorithms are implemented as a set of library routines written in C, and linked to the OpenMP-Fortran kernel programs. We put the corresponding labeling functions immediately after each 'C$OMP END PARALLEL DO' directive and each DO statement which follows 'C$OMP PARALLEL DO' directive. Each protocol function is located immediately after every statement that has the corresponding type of accesses to read or write to a shared variable. The data structures for the algorithms are instrumented in two types: a globally-shared access history for the protocol functions at the

Table 1. The Result using One-way Nested Kernel Programs (unit: second)

Kernel			Labeling		Protocol		Total	
N	t	Version	Original	T-BD	NR	T-BD	NR	T-BD	NR
	10	$_3P_{10}$	0.01	0.00	0.00	0.04	0.01	0.06	0.02
3	15	$_3P_{15}$	0.01	0.01	0.01	0.13	0.03	0.15	0.05
	20	$_3P_{20}$	0.01	0.02	0.01	0.31	0.09	0.34	0.11
	10	$_4P_{10}$	0.01	0.04	0.02	0.44	0.11	0.49	0.14
4	15	$_4P_{15}$	0.03	0.15	0.10	2.12	0.55	2.30	0.68
	20	$_4P_{20}$	0.07	0.46	0.31	6.48	1.69	7.01	2.07
	10	$_5P_{10}$	0.06	0.83	0.20	4.85	1.17	5.74	1.43
5	15	$_5P_{15}$	0.31	5.98	1.53	34.57	8.22	40.86	10.06
	20	$_5P_{20}$	1.13	24.53	6.36	140.95	33.45	166.61	40.94

Table 2. The Result using Two-way Nested Kernel Programs (unit: second)

Kernel			Labeling		Protocol		Total	
N	t	Version	Original	T-BD	NR	T-BD	NR	T-BD	NR
	10	$_3P_{10}$	0.01	0.03	0.02	0.33	0.08	0.37	0.11
3	15	$_3P_{15}$	0.02	0.07	0.05	1.08	0.28	1.17	0.35
	20	$_3P_{20}$	0.03	0.19	0.11	2.49	0.70	2.71	0.84
	10	$_4P_{10}$	0.07	0.48	0.33	7.23	1.84	7.78	2.24
4	15	$_4P_{15}$	0.32	2.34	1.60	34.87	8.47	37.53	10.39
	20	$_4P_{20}$	0.89	7.35	5.00	107.57	26.11	115.81	32.00
	10	$_5P_{10}$	1.45	9.53	6.45	161.42	35.57	172.40	43.47
5	15	$_5P_{15}$	9.02	71.02	48.04	1170.14	256.03	1250.18	313.09
	20	$_5P_{20}$	35.22	295.39	199.49	4818.61	1050.15	5149.22	1284.86

beginning of program, and the private variables for labeling functions which is instrumented as a `PRIVATE` clause in every 'C$OMP END PARALLEL DO' directive.

To compare the execution times of the two labeling schemes, we monitored the programs sequentially, because the comparison requires to measure the net execution time of labeling algorithm in the monitored execution. For the sequential monitoring, we set an OpenMP environment variable to execute the kernel program in one thread. Using the *time* command, each kernel program is measured in three phases: the *unmonitored phase* to measure the original execution time of the program, the *partially-monitored phase* to measure the time to execute only the labeling functions without instrumenting the protocol functions, and the *fully-monitored phase* to measure the total time to detect races. Then we calculated the protocol time by subtracting the partially-monitored time from the fully-monitored time.

The empirical results in Table 1 and 2 show that NR Labeling is more efficient than T-BD Labeling by at least 1.5 times in generating labels, by at least 3.5 times in using the labels to detect races, and by at least 3.0 times in total monitoring with the two sets of kernel programs. We argue that the differences reflect the time complexities of algorithms. The difference in the labeling times stems from the number of **for**-loops which consume $O(N)$ time in the worst-case in each labeling function for a thread started from fork operation: two loops in T-BD-Fork(), and one loop in NR-Fork(). And the difference in protocol time stems from the **while**-loops which compares the logical concurrency and the left-of relation between two threads. T-BD-Ordered() has a loop that consumes $O(N)$ time, but NR-Ordered() has a loop in its binary search that consumes only $O(\log_2 N)$, in the worst-case. And, T-BD-LeftOf() also has a loop that consumes $O(N)$ time in the worst-case, but NR-LeftOf() consumes only a constant time.

6   Conclusion

To compare the time efficiency of two scalable labeling schemes, BD Labeling and NR Labeling, we modified BD Labeling to be more efficient by making it generate only one label for each thread, and call it *Thread-based-BD (T-BD) Labeling*. Then, we empirically compared the actual efficiencies of these two scalable schemes using a set of OpenMP kernel programs with nested parallelism. We monitored the kernel benchmark programs sequentially, because the comparison requires to measure the net execution time of each labeling scheme in a monitored execution. The empirical results show that NR Labeling is more efficient than T-BD Labeling by at least 1.5 times in generating labels, by at least 3.5 times in using the labels to detect races, and by at least 3.0 times in total monitoring with the kernel programs.

References

1. Audenaert, K., "*Maintaining Concurrency Information for On-the-fly Data Race Detection*, Parallel Computing 97, pp. 1-8, North-Holland, Sept. 1997.
2. Dagum, L., and R. Menon, "*OpenMP: An Industry-Standard API for Shared-Memory Programming*," Computational Science and Engineering, 5(1): 46-55, IEEE, January-March 1998.
3. Dinning, A., and E. Schonberg, "*An Empirical Comparison of Monitoring Algorithms for Access Anomaly Detection*," 2nd Symp. on Principles and Practice of Parallel Programming, pp. 1-10, ACM, March 1990.
4. Jun, Y., and K. Koh, "*On-the-fly Detection of Access Anomalies in Nested Parallel Loops*," 3rd Workshop on Parallel and Distributed Debugging, pp. 107-117, ACM, May, 1993.
5. Lamport, L., "*Time, Clocks, and the Ordering of Events in a Distributed System*," Communications of the ACM, pp. 558-565, July 1978.
6. Mellor-Crummey, J., "*On-the-fly Detection of Data Races for Programs with Nested Fork-Join Parallelism*," Supercomputing '91, pp. 24-33, ACM/IEEE, Nov. 1991.
7. OpenMP Architecture Review Board, *OpenMP Fortran Application Program Interface*, version 2.0, Nov. 2000.
8. Sato, M., S. Satoh, K. Kusano, and Y. Tanaka, "*Design of OpenMP Compiler for an SMP Cluster*," 1st European Workshop on OpenMP, Lund, Sweden, Sept. 1999.

Debugging OpenMP Programs Using Event Manipulation

Rene Kobler, Dieter Kranzlmüller, and Jens Volkert

GUP Linz, Johannes Kepler University Linz,
Altenbergerstr. 69, A-4040 Linz, Austria/Europe,
{kranzlmueller|kobler}@gup.uni-linz.ac.at,
http://www.gup.uni-linz.ac.at/

Abstract. Debugging nondeterministic parallel programs is accepted as one of the harder problems of software engineering. One source of nondeterminsm are semaphores used to establish and control critical sections. As some threads compete for a semaphore, the point of time by which a specific thread locks a specific semaphore is not determined and may change during subsequent executions. A technique for debugging programs containing such race conditions is event manipulation, which allows the user to investigate the effects of different ordering in accesses to semaphores during subsequent re-executions. This allows to detect hidden errors, that may otherwise occur only sporadically. The technique described in this paper targets at OpenMP programs, and is therefore the first approach to perform event manipulation on shared memory applications.

1 Introduction

Shared memory parallel programming is an important paradigm for application development on high performance computing systems. Firstly, with OpenMP [1] there exists a standard allowing to write shared memory programs which are portable across most existing parallel architectures. Secondly, the growing spread of low-cost computing clusters constructed of commodity SMP nodes reaches more and more customers. Thirdly, even large-scale massively parallel machines like ASCI White[1] incorporate shared memory programming by pushing a mixed mode between MPI and OpenMP. Due to these reasons, it seems justified and necessary to provide powerful shared memory program development tools for software engineers.

As commercial available OpenMP debugging tools offer no functionality to deal with nondeterminism, our intention was to develop a method to handle nondeterministic OpenMP programs. In this paper, we describe a technique for supporting the user during debugging of nondeterministic OpenMP programs. Nondeterminism is observed, whenever subsequent program executions reveal different results although the same input is provided [7]. At present, we consider

---

[1] http://www.llnl.gov/asci/platforms/white/

R. Eigenmann and M.J. Voss (Eds.): WOMPAT 2001, LNCS 2104, pp. 81–89, 2001.

the order of accesses to shared data spaces, which are encapsulated in OpenMP lock and unlock operations. During an initial program execution, the observed access order is stored in tracefiles. Based on the stored data, an event graph is constructed, and the user can choose a different access order by applying event manipulation. Of course, only changes that are actually possible can be applied. The effects of these changes and their influence on a program's results can afterwards be evaluated by initiating an artificial program re-execution.

The next section offers some basic definitions on nondeterminism in parallel programs. Afterwards, the basics of the event manipulation strategy are introduced, before describing each step of this approach with an example of a nondeterministic OpenMP program.

2   Nondeterminism and Race Conditions

In parallel programs, nondeterministic program behavior is introduced either through special functions (e.g. the random number generator) or as a side effect of process interaction [3]. For process interaction in shared memory programs, two types of race conditions can be distinguished [10]:

- *data races*: if several threads can access shared memory and the accesses depend on the relative speed of the threads, different results can occur in successive program runs.
- *synchronization races*: when several threads perform synchronization operations on the same object, the relative speed of the threads may lead to different access ordering.

Data races are often not intended by the programmer, and therefore represent errors per se that can be removed immediately [10]. On contrary, synchronization races occur due to the user's placement of synchronization operations. An example is offered by OpenMP with the possibility of allocating semaphores and using corresponding lock and unlock operations. Figure 1 shows a synchronization race between two threads (T1, T2). The brackets indicate the lock and unlock operations for a selected semaphore operation, which is used to determine the value of variable A. As shown in Figure 1, two possible execution orders can occur, with different results in variable A depending on which process is allowed to perform its lock operation first.

Due to such race conditions, several difficulties occur during debugging [4]:

- The *irreproducibility effect*: given a nondeterministic program it cannot be guaranteed, that an observed execution can be reproduced during debugging.
- The *completeness problem*: obtaining all possible results of a nondeterministic program may be impossible, no matter how many executions are initiated.
- The *probe effect*: due to the overhead of debugging, a nondeterministic program may yield different results compared to an execution without debugger.

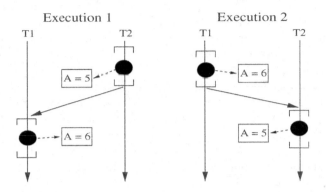

Fig. 1. Different execution sequence due to synchronization race.

The irreproducibility effect is addressed with so-called "record&replay" techniques, which operate in two phases [2]. After instrumenting a program, critical events are recorded during an initial executions. The recorded data can afterwards be used during arbitrary replay phases to establish the same order at the critical events, and thus provides an equivalent execution [6]. Record&replay approaches for shared memory parallel programs are discussed in [10,8].

In contrast to the irreproducibility effect, there are currently no solutions addressing the completeness problem and only a few approaches addressing the probe effect. In both cases, the main problem is to address the changes in access ordering. This problem is approached by our event manipulation approach as described below.

3   Event Manipulation Strategy

In order to describe our event manipulation approach, we should briefly offer some definitions concerning the underlying event graph model [4]:

Definition 1. *An event graph is a directed graph $G = (E, \rightarrow)$, where E is the non-empty set of events e_p^i of G, while $\rightarrow$ is the "happened before"-relation connecting events, such that $e_p^i \rightarrow e_q^j$ means, that there is an edge from event e_p^i to event e_q^j in G with the "tail" at event e_p^i and the "head" at event e_q^j.*

The events in such a graph are state changes occurring during program execution. An event definition is given in [9]:

Definition 2. *An event e_p^i is defined as an action without duration that takes place at a specific point in time i and changes the state of a process/thread p.*

The relation connecting these events is the so-called happened before relation, which is often used to establish a connection between corresponding events in concurrent systems [5]:

Definition 3. *The "happened before" relation denoted as "$\rightarrow$" on a set of events in G is the smallest transitive, irreflexive relation satisfying the following two conditions for arbitrary events e_p^i and e_q^j:*

1. *If e_p^i occurs before e_q^j on the same thread ($p = q$), then $e_p^i \rightarrow e_q^j$.*
2. *If e_p^i is the source of a communication or synchronization operation on thread p, and e_q^j is the corresponding target operation on thread q, then $e_p^i \rightarrow e_q^j$.*

The events in such a graph are state changes occurring during program execution [9]. At present, we are only interested in events generated by lock and unlock operations, which are responsible for synchronization races as described above.

The data for constructing an event graph model of a specific program run is generated during an initial execution. For that reason, a program has to be instrumented in order to observe the program's execution. Afterwards, the user can can perform nondeterminism analysis by performing the following steps [4]:

1. Selection of an arbitrary nondeterministic event
2. Evaluation of race candidates corresponding to the selected event
3. Event manipulation by exchanging the selected event with another candidate
4. Artificial replay to enforce the manipulated execution

4   Example

4.1   OpenMP Source Code

We describe this approach by means of a short OpenMP program example. In this example, computation is done by simply increasing an integer variable by 1 inside the critical section.

```
#include <omp.h>
#include <monitor.h>

#define SET(s,v) { omp_set_lock(&s);v++; omp_unset_lock(&s); }

omp_lock_t s1;
omp_lock_t s2;
omp_lock_t s3;

int v1 = 0, v2 = 0;

void main(int argc, char **argv) {
    omp_init_lock(&s1);
    omp_init_lock(&s2);
    omp_init_lock(&s3);
```

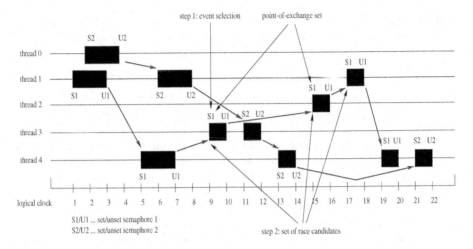

Fig. 2. Example event graph for nondeterminism analysis

```
#pragma omp parallel
{
    if (omp_get_thread_num() == 0) SET(s2,v2);
    else if (omp_get_thread_num() == 1) {
            SET(s1,v1); SET(s2,v2);
            SET(s1,v1);
        }
    else if (omp_get_thread_num() == 2) SET(s1,v1);
    else if (omp_get_thread_num() == 3) {
            SET(s1,v1); SET(s2,v2);
        }
    else if (omp_get_thread_num() == 4) {
            SET(s1,v1); SET(s2,v2);
            SET(s1,v1); SET(s2,v2);
        }
    }
    omp_destroy_lock(&s1);
    omp_destroy_lock(&s2);
    omp_destroy_lock(&s3);
}
```

A concrete event graph of an observed execution is shown in Figure 2. As events we have chosen the "omp_set_lock" and "omp_unset_lock" operations. For example, "S1" indicates an "omp_set_lock" operation on a arbitrary semaphore "semaphore 1", while "U1" describes the corresponding "omp_unset_lock" operation [1].

4.2 Evaluation of Race Candidates

For this example event graph, we apply the strategy defined above. According to step 1, we choose an arbitrary event for a nondeterministic lock operation, e.g. event $(S1, 9, 3)$, which means a set-event concerning semaphore 1 on thread 3 at logical clock 9. Assuming, that we want to investigate the program's behavior corresponding to the possible choices at this event, we proceed with step 2 of our strategy.

The main idea of step 2 is to identify all the possibilities, that may have occurred instead of event $(S1, 9, 3)$. For our example this means, that we have to identify all the lock operations, that may have occurred instead of our selected event. This means, that all events occurring concurrently to the selected event have to be considered.

The corresponding set of critical events is called race candidates (rc), which can be defined as follows:

$$rc(e^i_p) = \{e^i_p\} \cup \{\min(e^j_q) \mid \neg(e^j_q \to e^i_p) \forall 0 \le q < n, p \ne q\}$$

For our example of Figure 2, the following set of race candidates concerning event $(S1, 9, 3)$ is obtained:

$$rc((S1, 9, 3)) = \{(S1, 9, 3)\} \cup \{(S1, 15, 2), (S1, 17, 1), (S1, 19, 4)\}$$

Besides that, we need to verify whether there exist other operations, which interfere with the order of accesses to S1. In concrete, we need to identify those events on other semaphores, for which the following condition holds:

$$e^i_p \to e^k_q \wedge e^k_q \to e^j_q$$

For our example this means, that we must eliminate element $(S1, 19, 4)$ from the set of race candidates. This event cannot occur instead of event $(S1, 9, 3)$ due to its dependency on semaphore S2, which is accessed by thread 3 at event $(S2, 11, 3)$. The resulting set of race candidates is therefore:

$$rc((S1, 9, 3)) = \{(S1, 9, 3), (S1, 15, 2), (S1, 17, 1)\}$$

4.3 Event Manipulation

With the set of race candidates, we can investigate the following question:

What would have happened, if the nondeterministic choices would have been different from what has been observed?

For our example, we want to investigate, what would have happened, if any other member of the set of race candidates would have occurred instead of event $(S1, 9, 3)$. Together with this other member, the point-of-exchange (poe) can be defined as follows:

$$poe = \{e^i_p, e^j_q \in rc\}$$

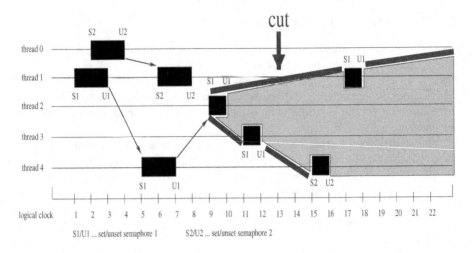

Fig. 3. Event graph after event manipulation and artificial replay

Therefore, we arbitrarily select event $(S1, 15, 2)$ to be exchanged with event $(S1, 9, 3)$ by specifying the following point-of-exchange:

$$poe = \{(S1, 9, 3), (S1, 15, 2)\}$$

Step 1 to 3 are shown in Figure 2 with annotations. With the *poe* defined, we can proceed to step 4. During step 4, the program is re-executed and the changes as defined by the *poe* are enforced. However, since the consequences of performing the event manipulation are unknown, we need to distinguish three phases during the artificial replay:

- Before the *poe*, the program is executed as previously observed. This resembles the behavior of traditional record&replay mechanisms.
- At the *poe*, the order of events is enforced as defined by the user.
- After the *poe*, the program needs to be executed without control. This resembles a program run without a replay mechanism.

4.4   Cut Placement

In order to identify the place, where our artificial replay mechanism needs to switch from the replay mode to an execution without constraints, we identify a so-called *cut*, which can be defined as follows:

$$cut(e_p^i) = poe \cup \{\min(e_r^k) \mid e_p^i \to e_r^k \forall 0 \leq r < n, r \neq p \neq q\}$$

The cut for our example of Figure 2 is shown in Figure 3. It consists of the following events:

$$cut((S1, 9, 3)) = \{(S1, 9, 3), (S1, 15, 2)\} \cup \{(S2, 13, 4), (S1, 17, 1)\}$$

The shaded region indicates the area, where the program's re-execution is carried out without the replay mechanism. No semaphore transitions are viewed in this region because the consequences of the event manipulation cannot be predicted. This is the reason why no transition from event $(U2, 8, 1)$ on thread 1 to the next set event on semaphore 2 is drawn.

The execution, that is revealed by the event manipulation, delivers the results with the revised event order. Therefore, it is possible that hidden errors or errors occurring only sporadically are detected.

5   Conclusions and Future Work

This paper describes the event manipulation approach for shared memory programs. At present, we support only the OpenMP set/unset operations, which are sufficient to show the usage of our technique. The technique has already been implemented in a first prototype tool. The tool is able to instrument arbitrary OpenMP programs (by using a C-preprocessor macros) and to generate corresponding tracefiles. Based on the tracefiles the four steps describe in Section 3 can be performed by the user. The missing part of this tool is a graphical user interface (GUI), which is currently being developed. A first view on this tool will be presented at the conference.

As a first extension we need to add other operations, e.g. the OpenMP test operation, which inspects the availability of a semaphore. At this point it is certainly necessary to apply the prototype tool to some real-world applications in order to identify remaining problems due to nondeterminism. In this context it is even imaginable, that the complete record&replay strategy as well as the event manipulation technique is included in an integrated program development environment. This would allow even inexperienced users to investigate nondeterministic parallel programs.

Another extension of our initial approach is to perform the event manipulation operation automatically. As shown in our example, we can choose several combinations of race candidates for the point-of-exchange. Consequently, there is more than one execution that can be derived from an initial program run with our event manipulation approach. If we are able to implement our strategy automatically, we will also solve the incompleteness problem, as well as problems with the probe effect due to event re-ordering.

Acknowledgements. Contributions to this work have been made by many people, most notable Michiel Ronsse, ELIS Dept., University Ghent, Belgium, who provided his knowledge about nondeterminism to bridge the gap between message-passing programs and shared-memory programs.

References

1. Chandra, R., Dagum, L., Kohr, D., Maydan, D., McDonald, J., Menon, R., "Parallel Programming in OpenMP", Academic Press, Morgan Kaufmann Publishers (2001)

2. Helmbold, D.P., McDowell, C.E., Wang, J.-Z., "Detecting Data Races by Analyzing Sequential Traces", Proc. HICCS-24, Hawaii Intl. Conference on System Sciences, Vol. 2, Hawaii, USA, pp. 408-417 (January 1991).

3. Kranzlmüller, D., Grabner, S., Volkert, J., "Debugging with the MAD Environment", Parallel Computing, Vol. 23, No. 1–2, pp. 199 217 (Apr. 1997).

4. Kranzlmüller, D., "Event Graph Analysis for Debugging Massively Parallel Programs", PhD Thesis, GUP Linz, Johannes Kepler University Linz, http://www.gup.uni-linz.ac.at/~dk/thesis (Sept. 2000).

5. Lamport, L., "Time, Clocks, and the Ordering of Events in a Distributed System", Communications of the ACM, pp. 558 - 565 (July 1978).

6. Leu, E., Schiper, A., and Zramdini, A., "Execution Replay on Distributed Memory Architectures", Proc. 2nd IEEE Symposium on Parallel & Distributed Processing, Dallas, TX, USA, pp. 106-112 (Dec. 1990).

7. Netzer, R.H.B., Miller, B.P., "What are Race Conditions? - Some Issues and Formalizations", ACM Letters on Programming Languages and Systems, Vol. 1, No. 1, pp. 74-88 (March 1992).

8. Netzer, R.H.B., "Optimal Tracing and Replay for Debugging Shared-Memory Parallel Programs", Proc. of the 3rd ACM/ONR Workshop on Parallel and Distributed Debugging, San Diego, CA, USA (May 1993).

9. van Rick, M., Tourancheau, B., "The Design of the General Parallel Monitoring System", Programming Environments for Parallel Computing, IFIP, North Holland, pp. 127-137 (1992).

10. Ronsse, M.A., De Bosschere, K., Chassin de Kergommeaux, J., "Execution Replay and Debugging", Proc. AADEBUG 2000, 4th Intl. Workshop on Automated Debugging, Munich, Germany, pp. 5-18 (August 2000).

The Application of POSIX Threads and OpenMP to the U.S. NRC Neutron Kinetics Code PARCS

D.J. Lee and T.J. Downar

School of Nuclear Engineering
Purdue University
W. Lafayette, IN 47907-1290
{lee100, downar}@purdue.edu

Abstract. POSIX Threads and OpenMP were used to implement parallelism in the nuclear reactor transient analysis code PARCS on multiprocessor SUN and SGI workstations. The achievable parallel performance for practical applications is compared for each of the code modules using POSIX threads and OpenMP. A detailed analysis of the cache misses was performed on the SGI to explain the observed performance. Considering the effort required for implementation, the directive based standard OpenMP appears to be the preferred choice for parallel programming on a shared memory address machine.

1 Introduction

The Purdue Advanced Reactor Core Simulator (PARCS) code[1] was developed by the United States Regulatory Commission for the safety analysis of nuclear reactors operating in the United States. The code solves the time-dependent Boltzmann transport equation for the neutron flux distribution in the nuclear reactor core. The code is written in FORTRAN and has been executed on a wide variety of platforms and operating systems. The code has been extensively benchmarked and results are documented in numerous reports[2] for a variety of Light Water Reactor (LWR) applications.

The computational burden in PARCS for solving practical reactor problems can be considerable even on the most efficient workstations. The execution time depends on the type of reactor transient simulated, but typically the code runs about an order of magnitude slower than real time. One of the primary objectives of the work performed here was to investigate multiprocessing as a means to reduce the computational burden with the ultimate goal of real time nuclear reactor simulation. The target platform for this work is a typical multithreaded workstation and therefore threading was the preferred parallel model. This paper will describe experience with using POSIX Thread[3] (Pthreads) and OpenMP[4] in achieving multiprocessing with the PARCS code on SUN and SGI workstations.

R. Eigenmann and M.J. Voss (Eds.): WOMPAT 2001, LNCS 2104, pp. 90–100, 2001.

The paper will first briefly discuss the physics and computational methods used in the PARCS code and then describe the code structure and approach used to achieve multiprocessing. Sections 3 and 4 will discuss the implementation of Pthreads and OpenMP in the PARCS code and analysis of the parallel performance for a practical reactor transient, respectively.

2   Physics and Computational Method

The solution of the Boltzmann transport equation for the neutron flux distribution in a nuclear reactor requires consideration of seven independent variables: neutron energy, three spatial dimensions, two angular dimensions, and time. Over the last several years, several approximations have proven successful in reducing the computational time and yet preserving the accuracy of flux predictions. For typical Light Water Reactor problems, two neutron energy groups and a first order angular flux approximation have proven to be effective. The neutron flux and fuel pin powers in the reactor can be predicted to within a few percent of measured data.

At any point in time, the resulting form of the equations is an elliptic PDE which can be discretized using conventional methods. The physical core model contains about 200 fuel assemblies, each of which contains about 250 fuel pins. Because the mean free path of the neutron is about the same as the distance between fuel pins, the resulting three-dimensional spatial mesh could be on the order of a million. However, innovative "multi-level" type methods have been developed over the years in which the pin by pin flux is solved only at the local level and the global problem is solved with mesh spacing the same size as the fuel assembly. The resulting local/global iteration introduces some natural parallelism since the local problems (NODAL module) can all be solved simultaneously. Innovative domain decomposition methods were also introduced to solve the global (CMFD module) problem. The Coarse Mesh Finite Difference (CMFD) solution is based on a Krylov semi-iterative method, Bi-Conjugate Gradient Stabilized (BICGSTAB), accelerated with a preconditioning scheme based on a blockwise incomplete LU factorization. Parallelism is achieved using an incomplete domain decomposition preconditioning method [1]. In addition to the Boltzmann transport equation, the single phase fluid dynamics and one-dimensional heat conduction equations are solved in PARCS to provide the temperature/fluid field in the core. This is performed in the T/H module of PARCS. The feedback to the neutron field equations is provided through the coefficients of the Boltzmann equation or "cross sections" and is performed in the XSEC module of PARCS.

Parallelism is achieved in PARCS by first dividing the code into the four basic modules which solve distinct aspects of the coupled neutron and temperature/fluid field calculation: CMFD, Nodal, T/H, and XSEC. The last three of these modules are easily parallelizable since all calculations are performed on a node by node basis and can be evenly assigned to separate threads. Parallelism of the CMFD calculation is more difficult but can be achieved by domain decomposition and incomplete LU preconditioner as noted above. In the analysis reported here domain decomposition is applied by dividing the reactor core into several axial planes. In the model used here the reactor core is divided into 18 axial planes. In some cases, load imbalance will contribute to the loss of parallel efficiency since the number of axial planes will not always be an even multiple of the number of processors.

3   Implementation of POSIX Threads and OpenMP

The target architecture for multiprocessing with PARCS was a shared memory address space machine since much of the parallelism was expected to be very fine grain. In general, the overhead for message passing parallelism is most suitable for applications which are predominantly coarse grain. In general, Pthreads and OpenMP are the standard techniques available for multithreaded programming. In the work here, both Pthreads and OpenMP were implemented on SUN and SGI workstations. Pthreads required considerably more effort than OpenMP to implement within the existing PARCS code. Considerable cautions had to be used with Pthreads to avoid such problems as race conditions. Conversely, the implementation of parallelism with OpenMP was considerably much simpler than Pthreads. It was simply a matter of inserting OpenMP directives for parallel regions and specifying which variables were shared and/or private.

3.1   POSIX Threads

Because Pthreads does not support an interface to FORTRAN, a mixed language programming was required in PARCS. A threads library was developed which is accessible to both FORTRAN and C sections of the code. It consists of a minimal set of thread functions for FORTRAN which were implemented by FORTRAN-to-C "wrappers". This library was named nuc_threads and provided the following four functions:

nuc_init(*ncpu)
This routine initializes mutex and condition variables. The initialization was called before any other nuc_threads subroutine and after defining the number of threads to be used.

nuc_frk(*func_name,*nuc_arg,*arg)
This routine creates the Pthreads. The arguments are the name of the function to be threaded, the number of arguments for the to-be-threaded function, as well as the actual arguments. The nuc_frk uses the Pthreads fucntion pthread_create.

nuc_bar(*iam)
This routine is used for synchronization. The nuc_bar uses a counter and the Pthreads routines pthread_cond_wait and pthread_cond_broadcast to implement synchronization.

nuc_gsum(*iam,*A,*globsum)
This routine is used to get a global sum of an array updated by each thread.

One approach to multithread a program is to fork each subroutine which contains a parallel region. The benefit of this approach is that it minimizes the global vs. local variable concerns. However, it introduces the possibility of performance loss due to the overhead of the nuc_frk function. This is particularly true if the subroutine is in a loop. It is very possible that the overhead associated with the use of nuc_frk could override any performance improvement by parallel computations. To avoid this problem and minimize the overhead in the application to PARCS, the main program was forked at the beginning of execution.

3.2 OpenMP

OpenMP is a standardized set of directives that can be used with FORTRAN, C, and C++ for programming on a shared address space machine. Compared with the Pthreads, the implementation is much easier due to the directive based nature of OpenMP. The nature of the parallelism in the OpenMP version PARCS (e.g. domain decomposition) was essentially identical to the parallelism in the Pthreads version implementation. However, there are some differences in the implementation such as threads being forked in each subroutine rather than in the main program as in the Pthreads implementation.

4 Application Results

The performance of Pthreads and OpenMP in PARCS was analyzed using a practical nuclear reactor transient benchmark problem. Prior to performing the practical transient analysis, however, it was helpful to first examine the performance of a simple FORTRAN program performing a matrix vector product. This analysis was used to gain some insight about the performance that was to be expected for the practical implementation in PARCS. The specifications for the two UNIX platforms used in this analysis are shown in Table 1.

4.1 Matrix-Vector Multiplication

The subroutine axb.f from the PARCS code which performs matrix-vector multiplication was parallelized using Pthreads and OpenMP. The size of vector used was 162KB (2x34x17x18x8) which is the same size as that of the benchmark problem presented in the next section. The timing results for the matrix vector application are summarized in Table 2. F90 version 6.1 is the latest available version of the SUN FORTRAN compiler. As shown in the results, the SUN FORTRAN compiler shows an unreasonable increase in execution time when using OPENMP for a single thread compared to unthreaded serial execution. This time increase appears to be attributable to the f90 dynamic memory allocation (DMA) module, since the threaded and serial time are similar when DMA is not used. However, DMA is implemented throughout PARCS and therefore must be utilized for multiprocessing. In contrast to the OpenMP performance, the execution time for the Pthreads implementation is very good on the SUN. As shown in the Table, the CPU time for serial and single thread execution are nearly identical and a near linear speedup (1.95) was observed for execution with 2 threads. The application of four and eight threads for the SUN platform was not practical since the SUN machines used here had only 2 CPUs.

Table 1. Specification of Machines

Platform	SUN ULTRA-80	SGI ORIGIN 2000
Number of CPUs	2	32
CPU Type	ULTRA SPARC II 450 MHz	MIPS R10000 250 MHz 4-way superscalar
L1 Cache	16 KB D-cache 16 KB I-cache Cache Line Size : 32bytes	32 KB D-cache 32 KB I-cache Cache Line Size : 32bytes
L2 Cache	4MB	4MB per CPU Cache Line Size : 128bytes
Main Memory	1GB	16GB
Compiler	SUN Workshop 6 -FORTRAN 90 6.1	MIPSpro Compiler 7.2.1 - FORTRAN 90

Table 2. Summary of Execution Time for MatVec Routine

Machine	Serial	OpenMP				Pthreads			
		1*	2	4	8	1	2	4	8
SUN	3.76**	23.43 (0.16)***	13.26 (0.28)	- -	- -	3.71 (1.02)	1.93 (1.95)	- -	- -
SGI	1.73	1.73 (1.00)	0.92 (1.89)	0.52 (3.30)	0.37 (4.72)	1.72 (1.01)	1.80 (0.96)	1.91 (0.91)	1.96 (0.88)

*   : Number of threads
**   : Execution time(milli-seconds)
*** : Speedup

On the SGI platform OpenMP shows good parallel performance, however parallel speedup could not be achieved with Pthreads since the scheduler assigned all threads to the same processor. The Pthreads standard is dependent on the vendor specific scheduler to assign threads to processors. Because this study was to compare the performance of Pthreads and OpenMP, no attempt was made to implement the SGI native threads library SPROC. It was observed that the difficulty in assigning threads to processors occurred only for mixed-language programming, i.e., this problem disappears for the C-language-only program using Pthreads.

For 2 threads execution on the SGI with OpenMP, a speedup of 1.89 was observed, which is comparable to the 1.95 speedup achieved on the SUN with Pthreads. There was a noticeable reduction in the parallel efficiency for this problem on larger numbers of processors (ie. 4 and 8 PEs). This is primarily because of load imbalance issues since the domain decomposition was restricted to the 18 axial planes which could not be evenly distributed to 4 and 8 PEs.

4.2 PARCS Application

The practical application with PARCS was to an international reactor transient benchmark problem[5]. An important transient in a nuclear reactor is the ejection of a control assembly from an initially critical core at hot, zero power conditions. There is a significant redistribution of power in the core and the possibility of a local energy deposition that could result in the melting of a fuel rod. The focus of the results here is on the neutron flux solution, but the transient also requires solution of the temperature/fluid field equations. The multithreaded solution of the OECD benchmark was performed with PARCS and as shown in Figure 1 below, all threaded versions provide the same result as the serial execution.

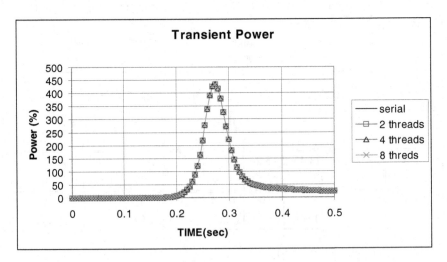

Fig. 1. Core Power versus Time for Serial and Multithreaded Code Execution

In Table 3 below, the execution time and parallel performance on the SUN machine are summarized for the Pthreads and OpenMP versions of the code. The number of updates in the table indicates the number of times each module must be executed in order to solve the benchmark problem. Similar to the results observed for the matrix-vector multiplication shown in Table 2, the OpenMP performance on the SUN is poor, whereas the Pthreads performance is reasonably good. The speedup achieved with the parallel Pthreads version is considerably different for each module. For example, the T/H and XSEC module show superlinear speedup while the CMFD and NODAL module speedup is only 1.77 and 1.78 respectively. As noted earlier, the T/H and XSEC modules are inherently parallel and therefore superlinear speedup is possible. Conversely, the solution of the elliptic PDE in the CMFD module uses domain decomposition methods and as indicated in the Table, there is a slight increase in the number of linear solutions required because of additional iterations. Also, the number of updates in the NODAL module increased due to the increased number of calls to the CMFD module. The increase of the number of updates for 2-threads case in CMFD and NODAL modules are small, just 2.5% and 6.5% respectively. Considering the naturally parallelizable nature of the NODAL module, the speedup of 1.78 is lower than expected even with the increase in the number of updates. A more detailed analysis was performed to understand these results from the standpoint of cache utilization and will be presented in the next section.

Table 3. Summary of Parallel Execution Times on the SUN

Module		Serial	OpenMP			Pthreads		
			1*	2	Speedup	1	2	Speedup
Time	CMFD	36.7	107.0	59.0	0.62	32.1	20.8	1.77
(sec)	Nodal	11.5	22.8	17.4	0.66	11.3	6.4	1.78
	T/H	29.6	32.6	19.1	1.56	27.9	14.5	2.04
	Xsec	7.6	47.5	24.3	0.31	7.1	3.7	2.04
	Total	85.4	209.8	119.8	0.71	78.5	45.5	1.88
# of	CMFD	445	445	456	-	445	456	-
Updates	Nodal	31	31	33	-	31	33	-
	T/H	216	216	216	-	216	216	-
	Xsec	225	225	226	-	225	226	-

* : Number of threads

The performance of PARCS on the SGI is summarized in Table 4. As indicated, the serial execution time on the SGI is reduced by almost 30% compared to the SUN. Similar to the Pthreads performance for the matrix vector multiplication shown in Table 2, it was not possible to schedule threads on the SGI and therefore no speedup was observed on the SGI with Pthreads. Conversely, the OpenMP performance on the SGI is very good. The speedup achieved with 2 threads (1.85) using OpenMP on the SGI is comparable to the 2-thread performance achieved with Pthreads on the SUN (1.88). In particular, the T/H and XSEC modules again show a superlinear speedup, whereas the CMFD and NODAL modules show a relatively low speedup. As will be discussed in the next section, this can be attributed in part to the efficiency of the cache utilization in each module. The speedups on 4 and 8 processors is poor, primarily because of problems balancing the load. As noted earlier, domain decomposition was implemented on a plane-wise basis and the 18 planes in the model are not evenly assignable to 4 and 8 processors.

Table 4. Summary of Execution Time on SGI

Module		Serial	OpenMP							Pthreads		
			1*	2	S**	4	S	8	S	1	2	S
Time	CMFD	19.8***	19.3	12.1	1.63	8.93	2.21	8.85	2.23	19.4	20.9	0.95
(sec)	Nodal	9.0	9.2	5.8	1.55	3.56	2.53	2.87	3.14	9.2	9.7	0.92
	T/H	26.6	25.3	12.3	2.17	8.92	2.99	7.14	3.73	25.2	25.5	1.05
	Xsec	4.8	4.4	2.4	2.01	1.37	3.53	1.11	4.35	4.8	5.0	0.97
	Total	60.2	58.1	32.6	1.85	22.8	2.64	20.0	3.02	58.6	61.1	0.99
# of	CMFD	445	445	456	-	497	-	565	-	445	456	-
Updates	Nodal	31	31	33	-	38	-	39	-	31	33	-
	T/H	216	216	216	-	216	-	217	-	216	216	-
	Xsec	225	225	226	-	228	-	227	-	225	226	-

```
*     : Number of threads
**    : Speedup
***   : Execution time (seconds)
```

4.3 Cache Performance Analysis

The processors on both of the platforms used in the analysis here are relatively fast, which suggests that data access times will be particularly important to improve overall code performance. Previous studies[6] have shown that cache utilization, particularly the L2 cache hit ratio, can be important for achieving good performance on modern workstations. Cache performance was analyzed for each PARCS module using the hardware counter library available on the SGI platform. As was shown in Table 4, the most time consuming part of the code is the T/H module, specifically the subroutine TRTH which solves the convection and heat transfer equations for the reactor coolant channel. The second most time consuming part of the code was the CMFD module, specifically the subroutine BICG which implements the Krylov method to solve the linear system. Therefore, BICG and TRTH were used to analyze cache performance for the CMFD and T/H modules, respectively. The cache performance of the other two modules, NODAL and XSEC, was analyzed for all the subroutines in the module. For the analysis shown here a consistent level 2 optimization was used for both serial and threaded versions of the code.

The data cache misses for each module are summarized in Table 5. The results were normalized to the serial misses as shown in Table 6. As shown in Table 6, the normalized L2 cache misses for 2 threads are 1.71 for XSEC, 1.66 for CMFD, and 1.60 for NODAL. This correlates reasonably well the speedups shown in Table 4 for XSEC, CMFD and NODAL of 2.01, 1.63 and 1.55, respectively. This suggests that for these three modules the overall performance is determined primarily by the L2 cache utilization. This is reasonable since, as shown in Table 7, an unsatisfied L2 cache miss can result in an order of magnitude penalty in machine cycles. The behavior of the T/H module does not appear to follow this trend since the normalized L2 data cache misses for 2 threads is very low (0.99) while the parallel performance is excellent (2.17). For the T/H module it appears that the L1 data cache plays a more important role. As shown in Table 5, the number of L1 data cache misses is about 20 times large than L2 cache data cache misses. As shown in Table 6, the L1 data cache misses for the threaded code are reduced by a factor of 4, which is very large compared with the other modules. Examination of the assembly code revealed some fundamental differences in the way the compiler treated the serial and threaded versions of the T/H subroutine. For example, the threaded version of the code had considerably more prefetch commands.

Table 5. Data Cache Misses

Module	Cache	Serial	OpenMP			
			1*	2	4	8
CMFD	L1	477,691	479,474	258,027	156,461	105,733
(BICG)	L2	28,242	29,650	17,007	11,751	9,309
Nodal	L1	857,744	853,866	444,849	249,507	160,699
	L2	54,163	55,534	33,846	19,016	12,848
T/H	L1	165,133	60,587	39,419	25,850	19,816
(TRTH)	L2	9,551	9,512	9,673	6,451	4,620
XSEC	L1	62,324	57,462	29,845	17,715	11,344
	L2	9,456	9,518	5,517	3,737	2,578

* : Number of threads

Table 6. Data Cache Misses (Normalized)

Module	Cache	Serial	OpenMP			
			1*	2	4	8
CMFD	L1	1.00	1.00	1.85	3.05	4.52
(BICG)	L2	1.00	0.95	1.66	2.40	3.03
Nodal	L1	1.00	1.00	1.93	3.44	5.34
	L2	1.00	0.98	1.60	2.85	4.22
T/H	L1	1.00	2.73	4.19	6.39	8.33
(TRTH)	L2	1.00	1.00	0.99	1.48	2.07
XSEC	L1	1.00	1.08	2.09	3.52	5.49
	L2	1.00	0.99	1.71	2.53	3.67

* : Number of threads

Table 7. Typical Memory Access Cycles

Memory Access Type	Cycles
L1 cache hit	2
L1 cache miss satisfied by L2 cache hit	8
L2 cache miss satisfied from main memory	75

In order to relate code performance to machine specific characteristics, the speedup was estimated using the data access times:

$$S \approx \frac{T_{total}^{serial}}{T_{total}^{2th}} \tag{1}$$

where

T_{total}^{serial} = Total data access time for serial execution

T_{total}^{2th} = Total data access time for 2 threads execution.

And the total data access time is approximated by L2 cache access time and main memory access time:

$$T_{total} \approx T_{L2} + T_{mem} = n_{L2} \cdot t_{L2} + n_{Mem} \cdot t_{Mem} \tag{2}$$

where

T_{L2} = Total L2 cache access time

T_{mem} = Total memory access time

n_{L2} = Number of L1 data cache misses satisfied by L2 cache hit

n_{Mem} = Number of L2 data cache misses satisfied from main memory

t_{L2} = L2 cache access time for 1 word

t_{Mem} = Main memory access time for 1 word.

Based on the above simple model and the data shown in Tables 5 and 7, the speedup for each module were estimated in Table 8. As indicated, the results are in reasonable agreement with the actual measured results shown in Table 4.

Table 8. Estimated 2-Thread Speedup Based on Data Cache Misses for OpenMP on the SGI

Module	Speedup	
	Measured	Predicted*
CMFD (BICG)	1.63	1.78
Nodal	1.55	1.80
T/H (TRTH)	2.17	2.04
XSEC	2.01	1.86

* : Predicted by Eq. (1)

5 Conclusions

Two parallel versions of the nuclear reactor transient analysis code PARCS were developed using Pthreads and OpenMP. The parallel performance was analyzed on SUN and SGI multiprocessors. Some unresolved issues remain regarding implementation of Pthreads on the SGI and OpenMP on the SUN. However, the overall performance of OpenMP on the SGI was comparable to Pthreads on the SUN. The performance varied depending on the type of calculations performed in each module. A simple predictive model was developed based on cache access times and the results agreed reasonably well with measured performance. Considering the effort required for implementation, the directive based standard OpenMP appears to be the preferred choice for parallel programming on a shared memory address machine. Future work will include the development of three-dimensional domain decomposition methods that will help alleviate load imbalance issues and improve the scalability of the code.

References

1. H.G. Joo and T.J. Downar, "An Incomplete Domain Decomposition Preconditioning Method for Nonlinear Nodal Kinetics Calculations," Nuc. Sci. Eng., 123, 403(1996)
2. H.G. Joo and T.J. Downar, "PARCS: Purdue Advanced Reactor Analysis Code", International ANS Reactor Physics Conference, Mito, Japan, 1996.
3. Bil Lewis and Daniel J. Berg. *THREADS PRIMER*. Californaia, USA: SunSoft Press, 1996.
4. OpenMP. A proposed industry standard api for shared memory programming. *http://www.openmp.org/*.
5. H. Finnemann, et al., "Results of LWR Core Transient Benchmarks," Proc. Intl. Conf. Math. And Supercomp. In Nuc. App., 2, p.243, Karlsruhe, Germany (April, 1993)
6. Q. Wang, "A Parallel Computing Model for the TRAC-M Code," M.S. Thesis, School of Nuclear Engineering, Purdue University, December, 1999.

Integrating OpenMP into Janus

Jens Gerlach, Zheng-Yu Jiang, and Hans-Werner Pohl

RWCP Parallel and Distributed Systems GMD Laboratory
GMD-FIRST, Kekuléstraße 7, 12489 Berlin, Germany
{jens,zhengyu,hans}@first.gmd.de

1   Introduction

OpenMP is an accepted portable programming model for shared-memory platforms. An advantage of OpenMP over MPI is that it offers a simple transition from sequential to parallel programs. Another question is, how OpenMP can be used in conjunction with higher level approaches to parallel programming.

In this paper, we report about our experiences regarding the integration of OpenMP into Janus—a parallel C++ template library for mesh and grid based scientific applications[3]. The issues emphasized here are *not* primarily to use OpenMP for the parallelization of a sequential program. The challenge is to map Janus' application-oriented and *explicit* description of communication onto OpenMP primitives in order to efficiently utilize the underlying shared memory.

Janus uses the ideas of *generic programming*[8] to identify abstractions and provides building blocks for complex parallel scientific applications. Janus clearly distinguishes between *spatial structures* (grids, meshes, or graphs) and the numerical values that are *associated* with them. The architecture and main template classes of Janus are briefly present in §2.

In §3, we discuss how OpenMP has been integrated into Janus components. In a subsequent section we discuss a parallel implementation of *Conway's Game of Life* that uses the `grid` and `stencil` templates from Janus. It turns out that OpenMP can be integrated smoothly into the Janus abstractions that were originally designed to hide the low level details of an underlying message passing system. Moreover, as we show in §5, no dramatic performance losses occur.

The test platforms for our two applications are both a for 4-processor Solaris shared multiprocessor system where we use the OpenMP compiler of KAI[6] and 16 node Linux PC Cluster featuring a fast MPI implementation. Finally, we summarize our experience and outline our future research.

2   The Janus Framework

The C++ template library *Janus*[3,4] provides basic building blocks for the implementation of mesh/grid-based scientific applications. The Janus components rest on a simple yet very expressive conceptual framework. Its basic idea is that there occur essentially two types of objects in scientific applications. These are, on the one hand, *spatial structures* which are, for example, grids, triangulations and graphs. On the other hand, we

R. Eigenmann and M.J. Voss (Eds.): WOMPAT 2001, LNCS 2104, pp. 101–114, 2001.

have (numerical) *data associated* with these spatial structures, such as grid functions, and (sparse) matrices. Fundamental for the Janus framework is the observation that the spatial structures are *prior to* and *more stable than* the associated data.

2.1 Generic Programming

David Musser defines *generic programming* as "programming with concepts" [8]. A Concept is set of type requirement. These requirements can be categorized into

- *interface requirements* that concern the syntax of involved types,
- *property requirements* that deal with semantic constraints.

A type that fulfills the requirements of a concept is called a *model* of that concept. Generic programming allows a grouping types that is independent of inheritance hierarchies. This avoidance of explicitly naming of a base interface is closely related to *structural subtyping*[11].

A well-known example of generic programming are the concepts and components of the C++ Standard Template Library[10] (STL) which provides many of the basic algorithms and data structures of computer science. Its basic concepts are Container and Iterator.

Algorithms of STL are written in terms of iterators, that is, independent of containers in data-structure neutral way. This decoupling of algorithms and containers reduces the code size of the STL drastically since a single template function can operate on different classes of containers. The *algorithm/data-structure interoperability* is a key aspect of STL's genericity. Two other aspects are *Function Objects* that can be supplied to containers and algorithms to adapt them to the need of a particular application, and *Element Type Parameterization*, that is, to plug basically every user-defined type into STL containers.

2.2 Concepts of Janus

The three basic concepts of the Janus framework are Domain, Relation, and Domain Function.

The role played by *iterators* in the STL is taken by the *position* of domain elements. They are the key for the *domain/relation interoperability* in Janus. They also decouple domain functions from the internal details of an underlying domain or relation.

Using *positions* instead of *iterators* appears at first sight less flexible. However, the *priority* and *relative stability* of the spatial structures, mentioned above, justifies this design decision.

Yet, the challenge is to express the relative stability in the context of irregular structures. Janus provides so-called *two phase* data structures with explicitly separated *insertion* and *retrieval* phases to solve this problem.

The Domain **Concept.** A domain is a finite set of elements that is given as a sequence. Each domain element has a unique (and fixed) position in its domain (its sequence index). The fixed one-to-one correspondence of domain elements and positions is crucial for Janus and a direct consequence of the *stability* of the spatial structures.

The basic requirements are similar to those of STL containers. A model of Domain must have nested types `value_type` and `size_type`. It must also provide methods `size()` to query its number of elements, `value_type operator[](size_type)` to return the ith element, and a method `size_type position(value_type)` to inquire the position of an element. These methods let domains appear like *searchable random access containers*.

Models of Two-Phase Domain must provide a method `insert(value_type)` and a method `freeze` that must be called *after* all elements have been inserted and *before* any element is accessed. Thus, `freeze` marks the transition from the insertion to the retrieval phase.

The Domain Function **Concept.** Data that are associated with domains or relations can exploit the fixed one-to-one correspondence of domain elements and their positions. The concept Domain Function describes data associated with a domain as a one-dimensional random access data-structure. It must have the same (fixed) length as the underlying domain or relation.

The Relation **Concept.** Domains have no visible structure. In order to describe dependences of domain elements there is the concept Relation. Relations between two domains are represented in terms of the positions of the domain elements and are described as *adjacency matrices*. In other words, a relation R between two domains X and Y is considered as the set of pairs of integers $R' = \{(i,j) \mid (x_i, y_j) \in R\}$.

The relations in scientific computations are often *sparse*. Therefore, the Relation concept provides operations that act on domain functions associated with its domains. These operations are generalizations of sparse matrix-vector multiplications and can be customized by function objects.

Its simplest form `pull` template function is in fact a sparse matrix-vector multiplication with the adjacency matrix. If b is a domain function on the domain Y then the result of *pulling b with R onto X* is a domain function a on X for which holds

$$a_i \longleftarrow a_i + \sum_{\{j \mid (i,j) \in R'\}} b_j \qquad \text{for all } i. \tag{1}$$

The *pull* and their transposed *push* operation are utilized to express (remote) data transfers between (distributed) domains and can be customized to perform user-defined operations on the gathered data.

The following table 1 gives an overview on the main *data transfer methods* provided by Janus relation classes. In particular the `pull_visitor` and `pull_matrix_visitor` templates are a very powerful means to specify complex user-defined operations. They have been inspired by the use of the *Visitor* Design Pattern[1] in the Boost Graph Library[7] (BGL).

Table 1. Main data transfer operations required by Relation

Name	Semantics for all positions i of X.
`pull(b, a)`	See Equation 1
`pull_fun(b, a, F)`	$F\left(\displaystyle\sum_{\{j\mid(i,j)\in R'\}} b_j, a_i\right)$
`pull_visitor(b, V)`	$V\left(\displaystyle\sum_{\{j\mid(i,j)\in R'\}} b_j, i\right)$
`pull_matrix(m, b, a)`	$a_i \longleftarrow \left(a_i + \displaystyle\sum_{\{j\mid(i,j)\in R'\}} m_{ij}b_j\right)$
`pull_matrix_fun(m, b, a, F)`	$F\left(\displaystyle\sum_{\{j\mid(i,j)\in R'\}} m_{ij}b_j, a_i\right)$
`pull_matrix_visitor(m, b, V)`	$V\left(\displaystyle\sum_{\{j\mid(i,j)\in R'\}} m_{ij}b_j, i\right)$

Note that all `pull` and related must be members of a relation class. The reason that we decided against globally defined template functions is that *pull* methods need a lot *low level state information* in order to be implemented efficiently. This holds in particular for the a parallel implementation which is the main focus of Janus.

This means that we heavily rely on *template member function* for the implementation of Janus.

2.3 Components of Janus

The core of Janus are six template classes that are shown in table 2. The first two classes `grid` and `domain` are models of Domain. The `grid` template can be used for a compact representation of rectangular grids. The `domain` template is a two-phase domain that can be used to represent irregular meshes.

The class `array` is a simple wrapper around the `std::vector` template. It provide overloaded versions of numerical operators and can query the size of an underlying domain or relation. It mainly exists for convenience—principle `std::vector` or even one-dimensional C-arrays fulfill the simple requirements of the concept Domain Function.

The remaining three classes are models of the concept Relation. They hide the details of three different sparse matrix formats and also provide methods for the convenient generation of the patterns related with these formats.

Janus does not provide many *proper* algorithms, i.e., algorithms that are implemented as functions and not as class methods. This is because the methods of the Relation classes are a very powerful mean to express complex data parallel operations. There are, however, parallel versions of some of the STL algorithms, for example, `count` and `accumulate`,

There is also a special class of *accumulator* algorithms in Janus that solve the problem of efficient assignment or values to a domain function on a *distributed* domain.

Table 2. Main template classes of Janus

`grid<N>`	An rectangular N-dimensional grid.
`domain<T>`	Indexed set of T objects.
`array<Dom,V>`	Domain function on an arbitrary Dom.
`stencil<N,Gen>`	General stencil between two N-dimensional grids.
`n_relation<X,Y,Gen>`	Regular sparse relation between two domains.
`relation<X,Y>`	General sparse relation between two domains.

3 Integration of OpenMP

The conceptual framework of Janus supports independent configurations of its components for *distributed* and *shared* memory architectures. As Figure 1 shows, basically all user-visible components are available for distributed and shared memory operation modes.

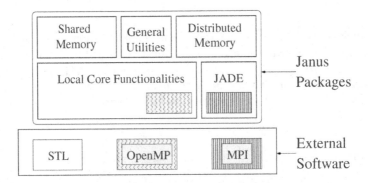

Fig. 1. The Janus Software Architecture

There are only slight differences regarding the interfaces and semantics. For example, mapping information can be specified for the domain template class when configured for distributed memory. Another difference concerns the size() of Domain. In case of a distributed configuration, it returns only the number of *local* elements. To inquire the number of all elements of a distributed domain the method total_size must be used.

The distributed subfamily of Janus rests on a (still experimental) port package JADE (*Janus Distributed Engine*). By default, JADE is implemented on top of MPI. We are also working on a port of Jade to the MTTL[5].

Both the shared and distributed memory abstraction share a considerable amount of *local core functionalities*. This includes basic sparse relation functionalities. This small subset is the place where OpenMP directives are mainly used.

To be more precise, we consider the general structure of the three models stencil, n_relation, and relation of the concept Relation. Each model R has an associated

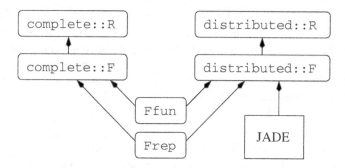

Fig. 2. Internal structure of the Janus relation classes. The arrows express *use* relationships. Only the `Ffun` module is annotated with OpenMP directives.

sparse (matrix) format F—see Figure 2. For example, the associated sparse format for `stencil` is derived from the diagonal sparse format DIA[9], whereas `relation` use the compressed row storage CRS[9] format. Different to conventional object-oriented design, a format F is split into and representation type `Frep` and function module `Ffun`. The module `Ffun` contains the basic operations to implement the `pull` methods of a relation (see table 1), and only this module contains OpenMP directives.

The following code fragment in Figure 3 shows the `pull` method from the function module of the DIA format. The structure of this method is very simple. I consists of a loop over a flag field that indicates whether the current position refers to an inner grid point where the stencil can safely applied.

```
1    template<typename Offset, typename T>
2    void pull(const Offset& offset, const size_vector& valid,
3                const T* s, T* t)
4    {
5        const size_t* flag = &valid[0];
6        int     fsize = valid.size(); // int is for guide++
7        int     i;
8   #ifdef _OPENMP
9   #pragma omp parallel shared(flag, fsize, s, t)
10       {
11  #pragma omp for private(i) schedule(runtime)
12  #endif  /* _OPENMP */
13           for(i = 0; i < fsize; i++)
14               if(flag[i]) dia_row::pull(i, offset, s, t);
15  #ifdef _OPENMP
16       }
17  #endif  /* _OPENMP */
18  }
```

Fig. 3. The Janus `dia::pull` method with OpenMP directives.

The actual operations are hidden in the small inline function `dia_row::pull`. The simple structure of the `dia::pull` makes it a perfect candidate to be generated as a wrapper around `dia::pull_row`. Regarding the many data transfer methods of a Janus relation class (see table 1), this would drastically simplify the tedious work of writing OpenMP directives. Generating the code was one reason to explicitly separate data representation and functionality into two different modules.

The class `Frep` and the module `Ffun` are used to implement

- `complete::F`—an implementation of F for shared memory,
- `distributed::F`—an implementation of F for distributed memory.

These classes are—as Figure 2 suggests—used by corresponding implementations of the Relation type R. We use conditional compilation[1] to map the names from the name spaces `complete` or `distributed` to our top level namespace `jns`.

This design has the advantage that OpenMP can be utilized not only for a shared memory computing platform but also for *distributed shared memory systems*.

4   An Application Case Study

Our goal is to integrate OpenMP into Janus in order to exploit OpenMP's ability to portably utilize the parallel paltforms with shared memory. Ideally, as mentioned above, we can hide OpenMP almost completely within the Janus abstractions. This is possible, if the Janus user avoids using loops that express data parallel operations. So the question is, whether Janus is *expressive enough* to support a wide range of data parallel scientific applications.

To preliminarily answer this question, we consider a simple scientific application: Conway's Game of Life. This is a well-known cellular automata simulation that involves rectangular grids and simple stencils to express data communication. We show how to write *parallel* implementations of this problems with the application-oriented Janus abstractions.

4.1   Conway's Game of Life

In *Life*, the evolution of a population of cells is considered. The underlying spatial structure is a rectangular grid (see Figure 4). A value of 1 represents a living cell (marked by ●), a value of 0 represents unoccupied grid points (not marked). The cells change their state in a lock step manner depending on the number of neighboring living cells and their own state. The neighborhood of a grid point is described by an 8-point stencil (see Figure 4). We apply the stencil only in the interior of the grid to exclude the issue of incomplete stencil environments at the boundary.

---

[1] The C preprocessor flags read _JANUS_SHARED_ and _JANUS_DISTRIBUTED_, respectively.

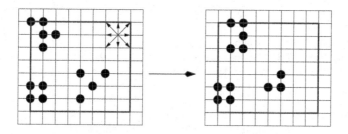

Fig. 4. Two successive generation in *Life*

4.2   A Janus Implementation of Life

The following code of Figure 5 shows an implementation of *Life* that uses the basic Janus classes `grid` and `stencil`. We comment the Janus constructs used in this program. Later we discuss alternatives that avoid the proliferation of OpenMP directives into the application program.

Lines 1 and 20 show the initialization and shutdown of Janus runtime environment. Depending on the actual configuration, these functions call their respective counter parts in MPI and OpenMP.

Lines 3 through 6 cover the initialization of the basic application parameters. The type nested type `grid<N>::value_type` (see §2.2) is a typedef for `svector<int,N>` which is a Janus utility for representing a container whose size is known at compile time. The variables `upper` and `lower` denote the upper right and lower left corner of the full grid. The variable `shift` is an offset for the respective entities of the inner grid.

Lines 7 and 8 declare the full grid object `outer` and its interior `inner`.

Line 9 shows the declaration and initialization of the 8-point stencil that describes the neighborhood of a grid point. For this the `stencil` template of Janus is used. Its first template parameter specifies the dimension of the involved grids. The second one is a *function object* (here `stencil8`) that gives a formal description of the actual stencil pattern. For *Life* this descriptions reads

$$(i,j) \longrightarrow \begin{cases} (i-1,j-1), & (i-1,j), & (i-1,j+1), & (i,j-1), \\ (i,j+1), & (i+1,j-1), & (i+1,j), & (i+1,j+1). \end{cases}$$

The definition of `stencil8` is shown in Figure 6. The `stencil8` function object is evaluated only in the constructor of `stencil` which transforms it into a very compact *sparse diagonal* representation.

Lines 10 through 12 show the declaration of two integer-valued domain function on the full grid. Here the `array` template of Janus is used. The state of the actual cell population is represented by the `state` domain function (remember: a value of 0 means *no cell* and 1 means *living cell*). The domain function `sum` will used later to hold for each grid point the number of living neighboring cells.

```
1   int main(int argc, char** argv) {
2         jns::initialize(argc, argv);
3         size_t iterations=500;
4         jns::grid<2>::value_type upper(1000, 1000);
5         jns::grid<2>::value_type lower(0,0);
6         jns::grid<2>::value_type shift(1,1);

7         jns::grid<2> outer(lower, upper);
8         jns::grid<2> inner(lower+shift, upper-shift);
9         jns::stencil<2,stencil8> neighbors(inner, outer);

10        jns::array<jns::grid<2>,int> state(outer.size());
11        // initialization of state not shown
12        jns::array<jns::grid<2>,int> sum(outer.size());
13        for(size_t i = 0; i < iterations; i++) {
14            neighbors.pull(state, sum);
15            for(size_t j = 0; j < outer.size(); j++)
16                evaluate(sum, state);
17        }
18        size_t c = jns::count(state, 1);
19        cout << c << " of " << outer.total_size() << endl;
20        jns::finalize();
21  }
```

Fig. 5. A straight-forward Janus implementation of *Life*

Lines 13 through 17 represent the main loop of the simulation. In each simulation step, at first, the number of living neighboring cells is counted for each grid point. For this the pull template function of stencil (see equation 1 of §2.2) is applied to the state domain function. In the subsequent loop, the number of living neighboring cells is evaluated and the state in each grid point is updated. The implementation of the hereby used inline function evaluate is given in Figure 7.

The remaining lines of the *Life* program in Figure 5 show how using the count algorithm of Janus the final number of living cells can be determined (in parallel). Note that in line 19 the method total_size must be called to determine the number of *all* elements of a (potentially distributed) domain (see also §3).

4.3   Discussion and a First Improvement

With respect to OpenMP, the two lines 15 and 16 are the biggest problem of the implementation shown in Figure 5. Putting the OpenMP directives into the application program is a far from optimal solution. Note that this is not an issue, when Janus is configured to solely use MPI since then this loop would be executed as part of a parallel processes.

```
1   struct stencil8 {
2        typedef jns::svector<int,2>           argument_type;
3        typedef jns::svector<argument_type,8> result_type;
4        result_type operator()(argument_type v) const {
5            result_type r(v);  // copy v
6            size_t k = 0;
7            for(int i = -1; i < 2; i++)
8                for(int j = -1; j < 2; j++)
9                    if((i != 0) || (j != 0))
10                       r[k++] += argument_type(i,j);
11           return r;
12       }
13   };
```

Fig. 6. Generator definition of a two-dimensional 8-point stencil

```
1   inline void evaluate(int& sum, int& state) {
2     if(sum == 3) state = 1;
3     else if ((sum == 2) && state) state = 1;
4           else state = 0;
5     sum = 0;
6   }
```

Fig. 7. Implementation of the evaluate function. Note that this function also resets the sum value to 0.

A straight-forward solution to this problem can be provide by a generic wrapper for simple loop expressions that hides the OpenMP directives. In fact, Janus provide the template function for_visitor that does exactly this.

In order to understand how this template function is used we look at the following code fragment in Figure 8.

4.4   A Second Improvement: Combining pull and Visitors

In the case of *Life*, we can add a further improvement by applying the visitor function object *while* executing pull. For this, Janus relations provide the template member function pull_visitor (see table 1). The big advantage is that *only one loop* must be executed during each iteration step.

The code fragment in Figure 10 replaces the lines 12 through 17 of example in figure 5.

5   Performance Measurements

We measured the time for the code fragment shown in Figure 10 because it is the fastest of the three Janus implementations (see also the Figures 5 and 8). We also wrote a

```
1        evaluate_visitor eval(&sum[0], &state[0]);
2        for(size_t i = 0; i < final; i++) {
3            neighbors.pull(state, sum);
4            jns::for_visitor(outer, eval);
5        }
```

Fig. 8. Using Janus for_visitor in *Life*. The five code lines replace the lines 13 through 17 of the original implementation in Figure 5. Instead of explicitly iterating over the outer grid object. The function object eval is applied by jns::for_visitor at each position of outer. The type of eval is evaluate_visitor whose implementation is shown in Figure 9. This class has pointers to the data of the domain functions state and sum that are initialize in line 1 of Figure 8.

```
1   struct evaluate_visitor {
2       int* sum;
3       int* state;
4       evaluate_visitor(int* sm, int* st) : sum(sm), state(st) {}
5       void operator()(size_t i) {
6         if(sum[i] == 3)
7             state[i] = 1;
8         else if ((sum[i] == 2) && state[i]) state[i] = 1;
9             else    state[i] = 0;
10        sum[i] = 0;
11      }
12  };
```

Fig. 9. Implementation of evaluate_visitor. This function object has only one entry point, namely, operator() that expects a single size_t argument. This argument is the domain position to which the function object is applied. Domain functions that shall be visited (for reading or manipulation) must pass the starting address of their data to the visitor. Remember, it is a fundamental Janus requirement (§2.2) that the positions of a domain are in one-to-one correspondence with the indices of an associated domain functions.

straight-forward implementation of *Life* to compare this with our Janus implementation. A 1000×1000 grid was used and 500 iteration steps performed.

Since Janus can be configured to use OpenMP or MPI, we also provide performance numbers for our 16 node Linux cluster.

5.1   Janus and OpenMP on a Shared Memory System

Our primary test system for this paper was a small shared-memory system with four CPU (Sun E3000, UltraSPARC-I 168MHz).

We used KAI's GUIDE OpenMP compiler (version 3.9) with the options +K3 -xO4 -fast -xtarget=ultra. In order to compare overheads due to the use of OpenMP, we also used Sun's KCC C++ Compiler (version 3.4) with the same options as GUIDE. We also measured times for using the GNU C++ compiler (version 2.95.2, using the options -O3 -funroll-loops -fexpensive-optimizations.

```
1        jns::array<jns::grid<2>,int> new_state(outer.size());
2        for(size_t i = 0; i < final; i++) {
3            evaluate_pull_visitor eval(&state[0], &new_state[0]);
4            neighbors.pull_visitor(state, eval);
5            state.swap(new_state);
6        }
```

Fig. 10. Using `pull_visitor` in *Life*. First we note that instead of the domain function sum the domain function `new_state` is used. In line 3, the start addresses of `state` and `new_state` are bound to the visitor function object `eval`. The type of `eval` is `evaluate_pull_visitor` whose implementation is shown in Figure 11. After calling `pull_visitor` in line 4 the state of the next generation of cells is stored in the domain function `new_state`. We use the `swap` member function (line 5) to efficiently assign `new_state` to `state`.

```
1   struct evaluate_pull_visitor {
2        const int* state1;
3        int* state2;
4        evaluate_pull_visitor(const int* s1, int* s2) :
5                              state1(s1),state2(s2){}

6        void operator()(int sum, size_t i) {
7          if(sum == 3) state2[i] = 1;
8          else if ((sum == 2) && state1[i]) state2[i] = 1;
9              else                          state2[i] = 0;
10       }
11  };
```

Fig. 11. Implementation of `evaluate_pull_visitor`. This function object has the only entry point `operator()`. It expects *two* arguments: the first one is the value that represents the result of `pull` at i-th position of the first domain. The second one *is* exactly this position. As in the case of the visitor in Figure 9, domain functions that shall be visited must pass the starting address of their data to the visitor.

The pure C implementation is roughly 1.5 faster than the Janus implementation. The C implementation also scales better and achieves a speedup of 4.52 for four threads (compared to a speedup of 3.62 in case of Janus). These are probably cache effects that are clearer visible in the C implementation. Therefore, the results for four threads show that the C implementation is almost times two times faster than Janus.

5.2   Janus and MPI on a Distributed Memory System

Our computing platform for the distributed case is a Linux cluster consisting of 16 Athlons running at 550 MHz. The nodes are connected through a Myrinet network that is utilized through the BIP[2] implementation of MPI.

Table 4 shows that the Janus program delivers an acceptable *absolute* speedup of 12 when using 16 nodes.

Table 3. Results for OpenMP on a Sun system

	Janus Implementation		C Implementation	
Compiler (Threads)	Time in s	Speedup	Time in s	Speedup
GUIDE (1)	66.87	1.00	42.59	1.00
GUIDE (2)	34.68	1.93	18.74	2.27
GUIDE (3)	23.31	2.87	12.21	3.49
GUIDE (4)	18.48	3.62	9.43	4.52
KCC	55.66	–	34.07	–
g++	62.37	–	35.23	–

Table 4. Result for MPI on a Linux cluster

Nodes	sequential	1	2	4	8	16
Time in s		48.03	63.05	31.64	15.92	7.94	3.97
Speedup		1.00	0.76	1.52	3.02	6.05	12.10

6   Future Works

So far, we found our experience with integrating OpenMP into Janus quite encouraging. For this (simple) application we could achieve reasonable speedup both when using OpenMP and MPI. More important, we could *completely* hide OpenMP directives within Janus. A next step for us is to integrate OpenMP into the remaining Janus components (see table 2). An Janus implementation of the Bellman-Ford Shortest Graph algorithm showed that the Janus constructs are also expressive enough for more complex and in particular irregular applications.

The next but one step is then to exploit the easy configurability of Janus and *combining* OpenMP *and* MPI within our framework. This would make our library very attractive for SMP clusters. So far however, we do not have access to SMP clusters *with* OpenMP support. This highlights a general drawback of the OpenMP approach. OpenMP must be integrated into a C++ compiler (in contrast to MPI). This hinders its availability, in particular on high performance machines.

Janus can be downloaded from www.first.gmd.de/janus.

References

1. E. Gamma, R. Helm, R. Johnson, and J. Vlissides. *Design Patterns: Elements of Reusable Object-Oriented Software*. Addison-Wesley, 1995.
2. Patrick Geoffray. Home Page of BIP Project. http://lhpca.univ-lyon1.fr/index_bip.html.
3. J. Gerlach and P. Gottschling. A Generic C++ Framework for Parallel Mesh Based Scientific Applications. In *Sixth International Workshop on High-Level Parallel Programming Models and Supportive Environments*, Lecture Notes in Computer Science, San Francisco, California, USA, April 2001. Springer-Verlag. See also http://www.first.gmd.de/janus.
4. J. Gerlach, H.W. Pohl, U. Der, Z.Y. Jiang, and P. Gottschling. A Janus Tutorial. Technical Report TR-00-01, RWCP Parallel and Distributed Systems GMD Laboratory, 2000.

5. Y. Ishikawa. Multi Thread Template Library – MPC++ Version 2.0 Level 0 Document. Technical report, RWCP, September 1996. RWC-TR-96-012, http://www.rwcp.or.jp/lab/pdslab/mpc++/mpc++.html.

6. Kuck & Associates. *Home Page KAI.* http://www.kai.com.

7. L.Q. Lee, J. Siek, and A. Lumsdaine. *Home Page of the Boost Graph Library (BGL).* C++ Boost. See also http://www.boost.org.

8. D. R. Musser. Generic Programming. www.cs.rpi.edu/~musser/gp/index.html.

9. Netlib Repository at UTK and ORNL. *Document for the Basic Linear Algebra Subprograms (BLAS) Standard.* Working document of the BLAST Sparse Subcommittee, available on http://www.netlib.org/utk/papers/sparse.ps.

10. Alex Stepanov. Standard Template Library Programmer's Guide. http://www.sgi.com/Technology/STL.

11. C. Szyperski. *Component Software: Beyond Object-Oriented Programming.* Addison-Wesley, 1999.

A Study of Implicit Data Distribution Methods for OpenMP Using the SPEC Benchmarks

Dimitrios S. Nikolopoulos[1] and Eduard Ayguadé[2]

[1] Coordinated Science Lab
University of Illinois at Urbana-Champaign
1308 West Main Str., Urbana, IL, 61801
dsn@csrd.uiuc.edu
[2] Department d' Arquitectura de Computadors
Universitat Politecnica de Catalunya
c/Jordi Girona 1-3, 08034, Barcelona, Spain
eduard@ac.upc.es

Abstract. In contrast to the common belief that OpenMP requires data-parallel extensions to scale well on architectures with non-uniform memory access latency, recent work has shown that it is possible to develop OpenMP programs with good levels of memory access locality, without any extension of the OpenMP API. The vehicle for localizing memory accesses transparently to the programming model, is a runtime memory manager, which uses memory access tracing and dynamic page migration to implement automatic data distribution. This paper evaluates the effectiveness of using this runtime data distribution method in non embarrassingly parallel codes, such as the SPEC benchmarks. We investigate the extent up to which sophisticated management of physical memory in the runtime system can speedup programs for which the programmer has no knowledge of the memory access pattern. Our runtime memory management algorithms improve the speedup of five SPEC benchmarks by 20–25% on average. The speedups are close to the theoretical maximum speedups for the problem sizes used and they are obtained with a minimal programming effort of about a couple of hours per benchmark.

1 Introduction

There is an ongoing debate between developers and users of OpenMP, about how should OpenMP be extended to scale better on architectures with non-uniform memory access latency, such as ccNUMA multiprocessors and clusters of SMPs [2,3,11]. Most of the related proposals converge to the conclusion that OpenMP should be extended with data distribution directives similar to the ones used in data-parallel programming languages like HPF. This argument is counterweighted by a recent research outcome [8,10] which suggests that it is possible to replace manual data distribution with intelligent runtime memory management algorithms, which infer the memory access pattern of the program

R. Eigenmann and M.J. Voss (Eds.): WOMPAT 2001, LNCS 2104, pp. 115–129, 2001.

and the most appropriate placement of data from traces of memory accesses collected in hardware counters.

We have developed a runtime memory manager which localizes transparently the memory accesses of OpenMP programs on tightly-coupled ccNUMA multi-processors [10]. The memory manager utilizes snapshots of memory access traces and dynamic page migration, to perform data distribution in a manner that minimizes the latency of remote memory accesses. The distinguishing feature of our runtime system is that it works transparently to the programmer and requires no modifications to the OpenMP API. It requires merely a simple instrumentation pass by the OpenMP translator to activate the memory manager. The runtime system detects automatically the data segment of the program and applies page migration algorithms by scanning the data segment periodically. The algorithms can exploit feedback from compiler analysis for data distribution, however they are primarily designed to operate in a fully automated process. The programming overhead for using the algorithms amounts to recompiling and linking the program with the runtime system.

The design of our runtime data distribution method is consistent with the current design principles of OpenMP, which dictate that the implementation must hide the details of the parallel architecture and the underlying hardware/software interface from the programmer. We consider our work as part of a broader effort towards building parallel programming models that meet the requirements of code and performance portability simultaneously. The purpose is to render the average programmer able to rapidly develop efficient portable parallel code and minimize the *effort/speedup* ratio, using a combination of OpenMP and runtime techniques for performance tuning.

1.1 Motivation

We have successfully applied our runtime data distribution method in parallel programs where manual data distribution would otherwise be necessary to reduce the number of remote memory accesses [7,8,9,10]. So far, we have evaluated the performance of our runtime system (named *UPMlib* after user-level page migration) using the OpenMP implementations of the NAS benchmarks [5]. Using these codes as a starting point provided us with a handful of advantages. The NAS benchmarks are highly parallel codes with coarse-grain parallelism and high sustained efficiency on most scalable parallel architectures; they are already well-tuned by their providers, the OpenMP implementations in particular are tuned specifically for the Origin2000, encompassing sophisticated cache-conscious programming and NUMA-aware data placement; moreover, the best manual data distribution algorithms, as well as the best automatic data placement algorithms for the NAS benchmarks were *a-priori* known to us, therefore we had a well-defined basis for comparisons.

In this work, we report our experience from using our runtime data distribution method with floating-point benchmarks from the SPEC CPU2000 and the SPEChpc benchmark suites [12]. Compared to the NAS benchmarks, the SPEC

Table 1. Coverage of parallel code in the SPEC benchmarks with which we experimented.

Benchmark	Coverage	Maximum speedup (Amdahl's law)
swim	77%	4.34
mgrid	80%	5.00
applu	51%	2.04
equake	93%	14.28
climate	80%	5.00

codes present us with a different picture. The native SPEC CPU2000 benchmarks are not parallelized. A very short parallelization effort of about a couple of hours per benchmark indicated that the codes expose a degree of parallelism which is well below that of the NAS benchmarks. The coverage of parallel code (ratio of the execution time of parallel code over the execution time of the entire program on a single processor) in the benchmarks we parallelized averages 76% (see Table 1). Amdahl's law suggests that the theoretical maximum speedups that the benchmarks can attain is very limited (6.13 on average). Practically, the parallelized loops are too fine-grain to provide sizable speedups. The actual maximum speedup of the hand-parallelized codes on a 64-processor Origin2000 ranges between 1.3 and 10. Of more interest to our work, is the fact that we are not aware of what is the best data placement algorithm for these benchmarks on a NUMA system, in order for us to have a point of reference for the performance of *UPMlib*.

The objective of this study is to investigate the extent up to which memory access localization can improve the scalability of codes with the characteristics of the SPEC benchmarks, on medium-scale NUMA systems. An important property common to both the SPEC and the NAS benchmarks, is that the codes are iterative, i.e. they repeat the same piece of computation for a number of iterations that correspond to ticks of a virtual timer. This is the exact class of codes for which our runtime system is more effective in localizing memory accesses [9]. Given that our runtime system is already tested against manual data distribution using iterative codes with known best data placement algorithms [7], a study with the SPEC benchmarks can provide us with an indication of the importance of memory accesses locality, without the burden of investigating, implementing and testing explicit data distribution schemes for the benchmarks.

Our findings are summarized as follows: Proper relocation of pages for localizing memory accesses has a significant impact on the scalability of the SPEC benchmarks. Linking the codes with *UPMlib* yields a speedup of up to 45% over the minimum execution time of the unmodified OpenMP code and up to 50% over the execution time with the maximum number of processors on the system with which we experimented (a 64-processor SGI Origin2000). It can be argued that the same or even higher speedup could be obtained with manual data distribution or careful restructuring of the code to localize the memory access pattern of each processor. This argument is counterweighted by the fact that these transformations require substantially more programming effort and non-portable extensions to OpenMP.

A second outcome of our experiments is that a good fraction of the speedup obtained from our runtime page migration algorithms can be obtained with a simpler automatic page placement algorithm, which invalidates the pages that store the data accessed within parallel loops and maps them locally to each processor on a first-touch basis[1], the first time a parallel loop is executed. This mechanism resembles the generic first-touch page placement algorithm [6], but instead of being used from the beginning of execution and for the whole address space of the program, it is used right before the first iteration of the outer time-stepping loop that encapsulates the parallel computation and solely for regions of the address space which are likely to incur frequent remote memory accesses. The intuition behind the algorithm, is to place pages strictly according to the memory access pattern of the parallel code. In this way, automatic page placement is not biased by the effects of initialization, which may include sequential access of critical arrays (forcing the placement of relevant pages on a single node), or a parallel initialization phase the memory access pattern of which does not match the memory access pattern of the actual parallel computation.

The rest of this paper is organized as follows. Section 2 provides some background on the basic concepts of our runtime data distribution method. Section 3 outlines our methodology, Section 4 presents the results from our experiments and Section 5 concludes the paper.

2   Background

UPMlib uses dynamic page migration as a tool for implicit data distribution. The runtime system infers the memory access pattern of a program by taking snapshots of page reference counters [2]. The key idea of the memory management algorithms of UPMlib is to retrieve snapshots of reference traces that reflect accurately the memory access pattern of the whole program, or specific parts of the program for which data distribution is required to localize memory accesses. Using these snapshots, the runtime system distributes data transparently to the programmer, using a cost/benefit criterion based on the frequency and the latency of remote memory accesses to each page.

In iterative parallel codes (which constitute the majority of parallel codes in use today), implementing global data distribution with UPMlib is as simple as retrieving a snapshot of the memory access trace at the end of the first iteration of the parallel computation and applying the page migration criterion on this snapshot. It has been shown that this simple runtime technique performs at least

---

[1] The term *first-touch* refers to a page placement algorithm in which the processor that touches a page first maps the page to a local memory module.

[2] Currently, the system is implemented on the Origin2000, which collects per-node reference information for each page in hardware counters. The counters are copied back to memory by the operating system upon overflow.

Table 2. Parallelized loops in the SPEC CPU2000 benchmarks (subroutines enclosing the loops and loop labels given in parentheses).

171.swim	INITAL(50,60,70,75,86)
	CALC1(100,110,115)
	CALC2(200,210,215)
	CALC3Z(400)
	CALC3(300,320,325)
172.mgrid	PSINV(600)
	RESID(600)
	RPJ3(100)
	INTERP(400,800)
	ZERO3(100)
	ZRAN3(400)
173.applu	jacld(outer k-loop)
	jacu(outer k-loop)
	rhs(outer k- and j-loops)
183.equake	smvp(i-loop)
	all i-loops inside main time-stepping loop

as well and usually better than manual data distribution in coarse-grain, embarrassingly parallel codes like the NAS benchmarks [10], while it can be easily extended to implement data redistribution within iterations. Under certain circumstances, *UPMlib* outperforms data distribution, because the runtime library captures more accurate page reference information.

The main disadvantage of *UPMlib* is that it is prone to the overhead of the page migration algorithms, which require expensive operations such as data copying, TLB coherence maintenance and several system calls for accessing hardware counters. This disadvantage shows up in fine-grain codes, more specifically, when the execution time per iteration is in the order of a few tens of milliseconds or less. Note that this problem occurs in data distribution tools as well, it is not an inherent disadvantage of *UPMlib*. A second disadvantage in that the page migration algorithms of *UPMlib* require some form of repeatability in the memory access pattern. The algorithms can not handle adaptive programs, that is, codes that perform an unpredictable amount of computation in each time step. Both issues, i.e. the runtime overhead of *UPMlib* and adaptive programs are a subject of ongoing work. We attempt to address the former by parallelizing the page migration algorithms, via inlining the algorithms in the OpenMP threads. For the latter, we investigate statistical approaches for data distribution.

3   Methodology

We manually parallelized six benchmarks from the SPEC CPU2000 floating point suite, namely *swim, mgrid, equake, applu, apsi* and *lucas*. The results for *apsi, lucas* and *applu* are qualitatively similar. More specifically, the speedup of the benchmarks flattens beyond 2 processors, due to the very limited coverage of parallel code (around 50%). We omit the results from the executions of these benchmarks, since they do not appear to be of interest to our study.

We parallelized the most time-consuming loops in each benchmark. We did not attempt to exploit coarse-grain task-level parallelism, although some benchmarks might benefit from it [1]. Most of the loops that we parallelized are identified as parallel by the SGI MIPSpro compiler, however in some cases, most notably in *applu* and *equake*, we had to manually apply privatization, rewriting of loop indices and induction variable elimination. The parallelized loops for each benchmark are listed in Table 2. Although the effort spent in parallelizing the benchmarks was not major, we can state with some confidence that extracting more parallelism out of these benchmarks would probably require highly sophisticated interprocedural analysis and other parallelization techniques not found in most commercial compilers. It might also require an extended execution model that exploits multiple levels of task and data parallelism [1]. Even if these techniques are applied, there is no clear evidence of their effectiveness. We have also experimented with the SPEChpc mesoscale climate modeling code. Parts of this code are already parallelized with OpenMP directives. We ran this code as distributed in the SPEChpc96 suite.

We executed three versions of the benchmarks. The first version is the unmodified parallelized OpenMP code. This code is executed using *STATIC* loop scheduling for all parallel loops and the first-touch page placement algorithm of IRIX. The IRIX kernel-level page migration engine was disabled during the experiments. IRIX includes a competitive dynamic page migration engine, which is activated by setting the *_DSM_MIGRATION* environment variable. We have run numerous tests with the IRIX page migration engine and found no case were the engine could provide meaningful improvements. All benchmarks except *swim* were slowed down by the IRIX page migration engine (*equake* slowed down by as much as 17%). *Swim's* improvement was less than 2% on 32 processors. Note that all benchmarks except *swim* have no parallel initialization phase. Parallel initialization is helpful in certain cases, because if an automatic data placement algorithm is applied during initialization, data may be distributed quite effectively before the beginning of the main parallel computation.

The second version is the parallelized OpenMP code linked with *UPMlib*. *UPMlib* applies a competitive page migration criterion to pages accessed during the execution of parallel code at the end of the first and, if needed, subsequent iterations of the outer time-step loop [10]. During our experiments with the SPEC benchmarks, all migrated pages were identified after the execution of the first iteration and the runtime system's page migration engine was deactivated thereafter. The overhead of executing the page migration algorithm was therefore minimal. Note that although *UPMlib* minimizes its runtime overhead, the cost of page migration is still significant enough to account for, as shown in Section 4.

The third version of the benchmarks is produced from the OpenMP code with the following modification. The pages accessed during the execution of parallel code are invalidated with the *mprotect()* system call, right before the first iteration of the outer time-stepping loop. We install a handler for the SIGSEGV signal, which records the address of the faulting page and maps the page to a local frame (i.e. residing on the same node with the processor that incurs the fault) using *mmap()*. This modification implements a restricted form of first-touch page placement. In particular, pages are mapped locally to the processor that touches them first during the first executed instance of a parallel loop. This ensures that pages are placed on a first-touch basis, according to the page reference pattern of the parallelized part of the code, rather than the reference pattern of the program as a whole. Basically, the mechanism discards any inopportune placement of pages that might be performed from the operating system during the initialization phase. We applied one optimization in the algorithm, for situations where the memory access patterns of different parallel loops do not match. In these cases, the algorithm applies first-touch page placement during the most time consuming loops with the same access pattern. We applied the algorithms in all benchmarks except *swim*. *Swim* already includes a parallelized initialization phase, which performs implicitly the placement of pages that our modification would otherwise do during the first iteration of the time-stepping loop.

Our hardware platform is a 64-processor (32-node) SGI Origin2000, with MIPS R10k processors running at 250 MHz. Each processor has 32 Kbytes of split L1 cache and 4 Mbytes of unified L2 cache. The system has 12 Gbytes of DRAM memory. The experiments were conducted on an idle system. The reported execution times are medians of three runs. We used the train problem sizes. Although this was mandated by system administration restrictions in the time available for running experiments on an idle system, the selected problem sizes are coarse enough to indicate parallel performance trends[3]. The maximum number of processors used was 62, because we detected contention between the IRIX kernel and the application threads when the benchmarks used all 64 processors. The codes were compiled the with -O2 optimization level.

4   Results

Figure 1 illustrates the execution times of four benchmarks versus the number of processors. Note that the y-axes are drawn logarithmic and the minimum and maximum values of the axes are tuned according to each benchmark's parallel execution time, for the sake of readability. The labels correspond to the unmodified OpenMP code (*OpenMP*), the OpenMP code linked with *UPMlib* (*OpenMP+upmlib*) and the OpenMP code modified to use our optimized first-touch page placement algorithm (*OpenMP+opt_ft*).

We observe a clear trend in the results. *UPMlib* reduces the minimum execution time of the benchmarks, regardless of the number of processors on which this time is obtained, by up to 45% (see Table 3). On the maximum number of pro-

---

[3] The granularity of the train problem size is roughly equivalent to that of the Class A problem size of the NAS benchmarks. The execution time per iteration is in the order of several hundreds of milliseconds.

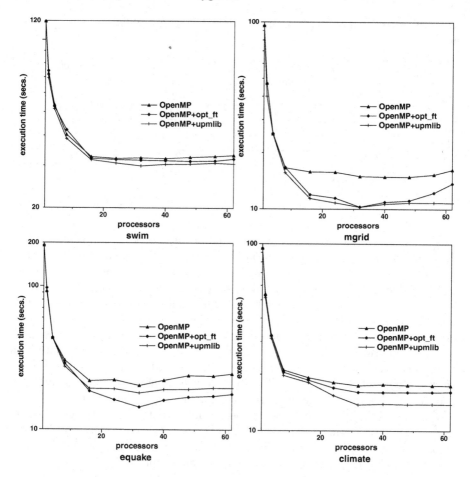

Fig. 1. Execution times.

cessors the improvement is up to 50%. On average, the margin of improvement is similar to that observed for the NAS benchmarks [7].

Figures 2 and 3 illustrate how *UPMlib* localizes memory accesses for higher performance. The charts show histograms of memory accesses from the executions of the benchmarks on 32 processors, divided into local (gray part) and remote (black part) memory accesses. The applied page migration algorithms have two important effects. First, they convert a significant fraction of remote memory accesses (in all cases more than 50%) into local memory accesses. On the Origin2000, this translates into a net saving of at least 100 ns. and up to 800 ns. per memory access [4]. The reductions in execution time are roughly proportional to the amount of remote memory accesses converted into local ones by *UPMlib*. The benchmarks with more remote memory accesses benefit more from dynamic page migration. *Mgrid* and *climate*, which average 1.5 and 6 million remote memory accesses per node, enjoy speedups of 45% and 26% respectively.

Table 3. Reduction of execution time (in percent) with *UPMlib* and our optimized first-touch algorithm.

	UPMlib		Optimized first touch	
	min. exec. time	exec. time on 62 proc.	min. exec. time	exec. time on 62 proc.
swim	-7.4	-8.7	-2.5	-3.7
mgrid	-44.8	-50.3	-44.5	-18.4
equake	-12.7	-25.7	-41.2	-38.4
climate	-26.4	-25.9	-8.7	-8.0
avg.	**-20.3**	**-27.7**	**-24.2**	**-17.1**
stdev	**16.7**	**32.4**	**29.7**	**20.5**

The second and apparently equally important effect of our runtime data distribution method is the alleviation of contention. The memory access traces reveal that all benchmarks have a highly unbalanced pattern of remote memory accesses. This means that a few nodes are accessed remotely significantly more frequently compared to the other nodes. The nodes that concentrate frequent remote memory accesses are likely to suffer from contention at their memory modules and network links. Contention is an important, yet underestimated effect on the performance of NUMA systems. On the Origin2000, the contention factor may account for as much as an additional 50 ns. per contended node per remote memory access [4]. *UPMlib* reduces contention by reducing and distributing evenly the remote memory accesses. The right charts in Figures 2 and 3 show that *UPMlib* achieves an almost perfectly balanced distribution of remote memory access in *mgrid* and *equake*. The remote memory accesses in *swim* are lightly unbalanced, however the low number of remote memory accesses per processor makes the contention effect almost imperceptible.

Intuitively, we expected somewhat more improvements, because unlike the OpenMP implementations of the NAS benchmarks [5], the SPEC codes are not tuned for efficient execution on a NUMA system. Practically, this effect does not show up in the experiments because the scalability of the SPEC benchmarks is primarily limited by the limited coverage of parallel code. A closer look at the results reveals that the benchmarks that have the higher speedup are also the ones that benefit more from the use of our page migration engine. Based on this observation, we speculate that extracting more parallelism out of the benchmarks is likely to make the impact of memory access localization more profound and the use of our runtime data distribution method vital. On the other hand, the magnitude of difference between the *OpenMP* and the *OpenMP+upmlib* versions indicates that the hardware of the Origin2000 is quite effective in reducing the impact of remote memory accesses, by guaranteeing a relatively low remote-to-local memory access latency ratio. We expect this trend to prevail in next-generation NUMA systems, which include full-fledged hardware mechanisms for reducing the impact of remote memory accesses, such as remote access caches and COMA protocols.

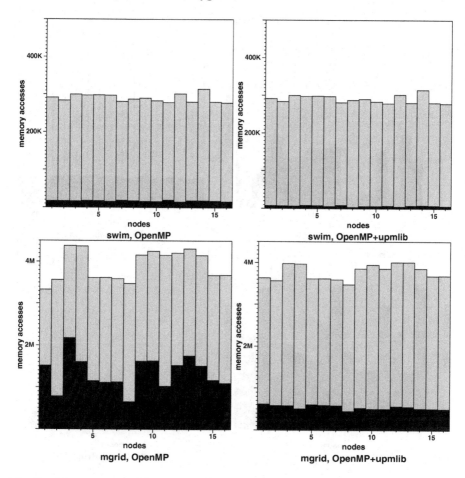

Fig. 2. Memory access histograms of *swim* and *mgrid* during their execution on 32 processors (16 nodes) of the Origin2000.

Interestingly, our optimized first-touch algorithm outperforms the native IRIX page placement algorithm. In two benchmarks, *swim* and *climate*, the improvements are considerably lower compared to those yielded by *UPMlib* (see Table 3). Figure 4 shows that in *climate*, the optimized first-touch algorithm has more remote memory accesses. The pattern of remote memory accesses is also highly unbalanced, which is a strong indicator of contention. The results show that first-touch is not the best choice of an automatic page placement algorithm for *climate*. We attempted to fix this problem using a round-robin page placement algorithm instead of first-touch without success. Round-robin performs slightly better than first-touch because it distributes better the remote accesses. However, the actual number of remote memory accesses is increased with round-robin. The memory access pattern of *climate* is both unbalanced

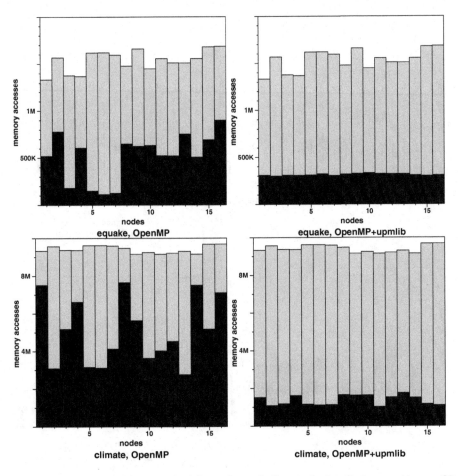

Fig. 3. Memory access histograms of *equake* and *climate* during their execution on 32 processors (16 nodes) of the Origin2000.

and irregular, therefore dynamic page migration is the best option for optimizing memory access locality.

Both the page migration algorithm and the optimized first-touch algorithm perform approximately the same in *mgrid*. In *equake*, the optimized first-touch algorithm outperforms the IRIX page placement algorithm by a significantly wider margin compared to our dynamic page migration engine. This result is somewhat surprising. Figure 4 shows that the optimized first-touch algorithm incurs less remote memory accesses than the page migration algorithm. We were not able to find a convincing explanation for this effect, other than that *UPMlib* relocates pages that concentrate frequent remote memory accesses later than the optimized first-touch algorithm. The reason which is more likely to explain the performance of the page migration engine in *equake* is the overhead of page migrations.

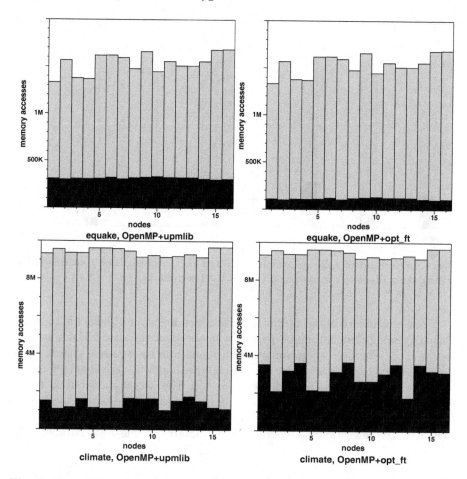

Fig. 4. Comparison of the memory access traces of the page migration engine and the optimized first-touch algorithm in *climate* and *equake*.

Figure 5 shows the relative overhead of the page migration algorithm during the executions of the benchmarks on 62 processors. *UPMlib* uses a thread that executes the page migration algorithms in parallel with the program. Although the overhead of page migration is overlapped, there is still an interference between the *UPMlib* thread and one or more *OpenMP* threads. We conservatively estimated this interference by measuring the CPU time spent by the *UPMlib* thread and assuming that 50% of this CPU time is spared from a thread of the program. The relative overhead of page migration in *equake* exceeds 20% of the execution time of the benchmark. As a comparison, the relative overhead of the optimized first-touch algorithm in *equake* is 4% of the total execution time (shown in the right chart of Figure 5). The overhead of page migration is noticeable in *swim* and *climate* as well.

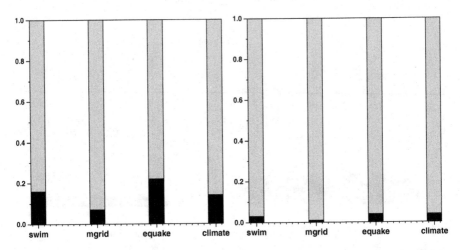

Fig. 5. Overhead of page migration (left) and the optimized first-touch algorithm (right), normalized to the execution time of the benchmarks.

The performance of our optimized first-touch algorithm brings life to an old idea suggesting that automatic page placement algorithms for NUMA systems can be more effective, when coupled with hints from the runtime system about the points of execution at which the algorithms should be activated (referred to as phase changes in the related literature [6]). The actual contribution to our OpenMP memory management framework is a low-cost, transparent mechanism for localizing memory accesses, which might prove of practical use in cases where dynamic page migration is vulnerable, most notably in fine-grain parallel codes [7].

5   Conclusions

This paper analyzed the performance of runtime memory management algorithms that localize the memory accesses of OpenMP programs on a NUMA system, using programs from the SPEC benchmark suites. The SPEC benchmarks are challenging for automatic memory management algorithms, because they are not embarrassingly parallel, neither are they tuned for efficient execution on NUMA systems. Linking the codes with our runtime system yielded a solid performance improvement of 20–25% on average. Similar or somewhat lower improvements were obtained from a simpler algorithm that places pages on a first-touch basis during the first invocation of each parallel loop. This mechanism may be valuable when the overhead of page migration is non-negligible. Overall, the results are consistent with a recently established trend that favors the use of intelligent runtime methods to scale OpenMP on clustered NUMA architectures, without modifying the appealing OpenMP API [9]. We plan to take several steps in this direction, hoping to secure performance portability with unmodified implementations of the OpenMP standard.

Acknowledgments. Jesús Labarta, Theodore Papatheodorou and Constantine Polychronopoulos have contributed valuable insight in earlier stages of this research. This work was supported by NSF Grant No. EIA-9975019 and the Spanish Ministry of Education Grant No. TIC98-511. The experiments were conducted with resources provided by the European Center for Parallelism of Barcelona (CEPBA).

References

1. E. Ayguadé, X. Martorell, J. Labarta, M. González, and N. Navarro. Exploiting Multiple Levels of Parallelism in OpenMP: A Case Study. In *Proc. of the 1999 International Conference on Parallel Processing (ICPP'99)*, pages 172–180, Aizu, Japan, August 1999.
2. S. Benkner and T. Brandes. Exploiting Data Locality on Scalable Shared Memory Machines with Data Parallel Programs. In *Proc. of the 6th International EuroPar Conference (EuroPar'2000)*, pages 647–657, Munich, Germany, August 2000.
3. J. Bircsak, P. Craig, R. Crowell, Z. Cvetanovic, J. Harris, C. Nelson, and C. Offner. Extending OpenMP for NUMA Machines. In *Proc. of the IEEE/ACM Supercomputing'2000: High Performance Networking and Computing Conference (SC'2000)*, Dallas, Texas, November 2000.
4. D. Lenoski C. Hristea and J. Keen. Measuring Memory Hierarchy Performance on Cache-Coherent Multiprocessors Using Microbenchmarks. In *Proc. of the ACM/IEEE Supercomputing'97: High Performance Networking and Computing Conference (SC'97)*, San Jose, California, November 1997.
5. H. Jin, M. Frumkin, and J. Yan. The OpenMP Implementation of the NAS Parallel Benchmarks and its Performance. Technical Report NAS-99-011, NASA Ames Research Center, October 1999.
6. M. Marchetti, L. Kontothanassis, R. Bianchini, and M. Scott. Using Simple Page Placement Schemes to Reduce the Cost of Cache Fills in Coherent Shared-Memory Systems. In *Proc. of the 9th IEEE International Parallel Processing Symposium (IPPS'95)*, pages 380–385, Santa Barbara, California, April 1995.
7. D. Nikolopoulos, E. Ayguadé, J. Labarta, T. Papatheodorou, and C. Polychronopoulos. The Trade-Off between Implicit and Explicit Data Distribution in Shared-Memory Programming Paradigms. In *Proc. of the 15th ACM International Conference on Supercomputing*, Sorrento, Italy, June 2001.
8. D. Nikolopoulos, T. Papatheodorou, C. Polychronopoulos, J. Labarta, and E. Ayguadé. A Transparent Runtime Data Distribution Engine for OpenMP. *Scientific Programming*, May 2001.
9. D. Nikolopoulos, T. Papatheodorou, C. Polychronopoulos, J. Labarta, and E. Ayguadé. Is Data Distribution Necessary in OpenMP ? In *Proc. of the IEEE/ACM Supercomputing'2000: High Performance Networking and Computing Conference (SC'2000)*, Dallas, Texas, November 2000.
10. D. Nikolopoulos, T. Papatheodorou, C. Polychronopoulos, J. Labarta, and E. Ayguadé. UPMlib: A Runtime System for Tuning the Memory Performance of OpenMP Programs on Scalable Shared-Memory Multiprocessors. In *Proc. of the 5th ACM Workshop on Languages, Compilers and Runtime Systems for Scalable Computers (LCR'2000)*, LNCS Vol. 1915, pages 85–99, Rochester, New York, May 2000.

11. V. Schuster and D. Miles. Distributed OpenMP, Extensions to OpenMP for SMP Clusters. In *Proc. of the Workshop on OpenMP Applications and Tools (WOM-PAT'2000)*, San Diego, California, July 2000.
12. Standard Performance Evaluation Corporation (SPEC). SPEC CPU2000 and SPEC hpc96 documentation. http://www.spec.org, accessed 2001.

OmniRPC: A Grid RPC Facility for Cluster and Global Computing in OpenMP
(Extended Abstract)

Mitsuhisa Sato[1], Motonari Hirano[2], Yoshio Tanaka[2], and Satoshi Sekiguchi[2]

[1] Real World Computing Partnership, Tsukuba, Japan
[2] Software Research Associates, Inc
[3] Electrotechnical Laboratory

Abstract. *Omni remote procedure call facility*, OmniRPC, is a thread-safe grid RPC facility for cluster and global computing environments. The remote libraries are implemented as executable programs in each remote computer, and OmniRPC automatically allocates remote library calls dynamically on appropriate remote computers to facilitate location transparency. We propose to use OpenMP as an easy-to-use and simple programming environment for the multi-threaded client of OmniRPC. We use the POSIX thread implementation of the Omni OpenMP compiler which allows multi-threaded execution of OpenMP programs by POSIX threads even in a single processor. Multiple outstanding requests of OmniRPC calls in OpenMP work-sharing construct are dispatched to different remote computers to exploit network-wide parallelism.

1 Introduction

In this paper, we propose a parallel programming model for cluster and global computing using OpenMP and a thread-safe remote procedure call facility, OmniRPC.

In recent years, two important computing platforms, a cluster of workstation/PC and a computational grid, have been gathering many interests in high performance network computing. Recent progress in microprocessors and interconnection networks motivates high performance computing using clusters out of commodity hardware. Advances in wide-area networking technology and infrastructure make it possible to construct large scale high-performance distributed computing environments, or computational grids that provide dependable, consistent and pervasive access to enormous computational resources.

Omni remote procedure call facility, OmniRPC, is a thread-safe implementation of Ninf RPC which is a grid RPC facility in a wide-area network. The remote libraries for OmniRPC are implemented as executable programs, and are registered in each remote computer. The OmniRPC programming interface is designed to be easy-to-use and familiar-looking for programmers of existing languages such as FORTRAN, C and C++, and is tailored for scientific computation. The user can call the remote libraries without any knowledge of the

R. Eigenmann and M.J. Voss (Eds.): WOMPAT 2001, LNCS 2104, pp. 130–136, 2001.

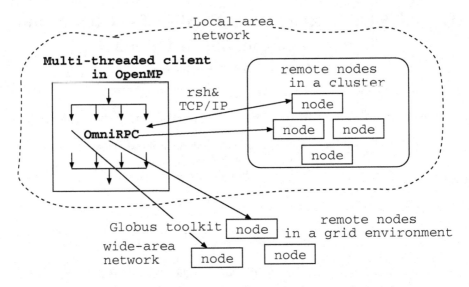

Fig. 1. OpenMP multi-threaded client and OmniRPCs

network programming, and easily convert his existing applications that already use popular numerical libraries such as LAPACK. A client can execute the time-consuming part of his program in multiple and heterogeneous remote computers, such as clusters and supercomputers, without any requirement for special hardware or operating systems. OmniRPC provides uniform access to a variety of remote computing resources.

At the beginning of execution, the initialization of OmniRPC collects the information about remote libraries registered in available remote computers. OmniRPC automatically allocates remote library calls dynamically on appropriate remote computers to facilitate location transparency. In order to support parallel programming, the multi-threaded client can issue multiple requests by OmniRPC simultaneously. Each outstanding request is dispatched to a different remote computer to exploit network-wide parallelism. Although the POSIX thread library can be used for programming multi-threaded clients, multi-threaded programming using thread library directly makes the client program complicated.

While the OpenMP Application Programming Interface (API) is proposed for parallel programming on shared-memory multiprocessors, OpenMP provides a multi-threaded programming model without the complexity of multi-threaded programming. We have developed Omni OpenMP compiler[4], which is a free and open-source, portable implementation of OpenMP. In a shared memory multiprocessor, threads in OpenMP are eventually bound to physical processors for efficient parallel execution. The POSIX thread implementation of the Omni OpenMP compiler allows multi-threaded execution of OpenMP programs by POSIX threads even in a single processor.

OpenMP provides an easy-to-use and simple programming environment for the multi-threaded client of OmniRPC. Figure 1 illustrates the OpenMP multi-threaded client and OmniRPCs. A typical application in cluster and grid environments is parametric execution which executes the same code with different input parameters. For this type of applications, we can use OpenMP parallel loop directives to execute OmniRPC calls in parallel for different remote computers.

For a computational grid environment, OmniRPC uses Globus toolkit[1] as a grid software infrastructure. Although a Globus implementation of MPICH, MPICH-G, can be used for parallel programming in Globus, message passing programming requires programmers to explicitly code the communication and makes writing parallel programs cumbersome. While our proposed model is limited to a master-slave model, it provides very simple parallel programming environment for a computational grid.

The parallel programming model with the OpenMP client of OmniRPC can be applied to other RPC facilities such as CORBA if these APIs are thread-safe. NetSolve[3] is a similar RPC facility to our OmniRPC and Ninf. It also provides a programming interface similar to ours and automatic load balancing mechanism by a agent. To our best knowledge, no experience of parallel programming with OpenMP is reported.

2   OmniRPC: A Thread-Safe Remote Procedure Call Facility

A client and the remote computational nodes which execute remote procedures may be connected via a local area network or over a wide-area network. A client and nodes may be heterogeneous: data in communication is translated into the common network data format.

The remote libraries are implemented as executable programs which contain network *stub* routine as its main routine, and registered in the *registry file* in each remote nodes. We call such executable programs *Ninf executables (programs)*. These stubs are generated from the interface descriptions by the Ninf IDL compiler.

In a client node, a user prepares his own *machine file* which contains the host names of available computation nodes. The OmniRPC initialization function, OmniRPC_init, reads registry files in the remote nodes to make the database which associates the entry names of remote functions with Ninf executables.

OmniRPC inherits its API and basic architecture from Ninf. OmniRPC_Call() is the sole client interface to call the remote library. In order to illustrate the programming interface with an example, let us consider a simple matrix multiply routine in C programs with the following interface:

```
double A[N][N],B[N][N],C[N][N];   /* declaration */
....
dmmul(A,B,C,N);   /* calls matrix multiply, C = A * B */
```

When the dmmul routine is available in a remote node, the client program can call the remote library using OmniRPC_Call, in the following manner:

```
OmniRPC_Call("dmmul",A,B,C,N); /* call remote library */
```

Here, dmmul is the entry name of library registered as a Ninf executable on a remote node, and A,B,C,N are the same arguments. As we see here, the client user only needs to specify the name of the function as if he were making a local function call; OmniRPC_Call() automatically determines the function arity and the type of each argument, appropriately marshals the arguments, makes the remote call to the remote node, obtains the results, places the results in the appropriate argument, and returns to the client. In this way, the OmniRPC is designed to give the users an illusion that arguments are shared between the client and the remote nodes.

To realize such simplicity in the client programming interface, a client remote function call obtains all the interface information regarding the called library function at runtime from the server. The interface information includes the number of parameters, these types and sizes and access mode of arguments (read/write). Using these informations, the RPC automatically performs argument marshaling, and generates the sequence of sending and receiving data from/to the nodes. This design is in contrast to traditional RPCs, where stub generation is done on the client side at compile time.

The interface to a remote function is described in Ninf IDL. For example, the interface description for the matrix multiply given above is:

```
Define dmmul(long mode_in int n, mode_in double A[n][n],
          mode_in double B[n][n], mode_out double C[n][n])
"... description ..."
Required "libxxx.o" /* specify library including this
                                           routine. */
Calls "C" dmmul(n,A,B,C);   /* Use C calling convention. */
```

where the *access specifiers* , mode_in and mode_out, specify whether the argument is read or written. To specify the size of each argument, the other in_mode arguments can be used to form a size expression. In this example, the value of n is referenced to calculate the size of the array arguments A, B, C. Since it is designed for numerical applications, the supported data type in Ninf IDL is tailored for such a purpose; for example, the data types are limited to scalars and their multi-dimensional arrays. The interface description is compiled by the *Ninf interface generator* to generate a stub program for each library function. The interface generator also automatically outputs a makefile with which the Ninf executables can be created by linking the stub programs and library functions.

To invoke a Ninf executable in a remote node, OmniRPC use the remote shell command "rsh" in a local area network and GRAM(Globus Resource Allocation Manager) API of Globus toolkit in a grid environment. The Ninf executable is invoked with the arguments of the client host name and port number for waiting the connection. For a grid environment, we designed OmniRPC on top of the Globus toolkit. The Globus I/O module is used in a grid environment instead of TCP/IP in local area network. The Globus also provides security and authentication by GSI (Globus Security Infrastructure).

To handle multiple outstanding RPC requests from a multi-threaded client, OmniRPC maintains the queue for outstanding remote procedure calls. OmniRPC_Call() enqueues the request for the remote call, and blocked for waiting the return from the remote call. The scheduler thread is created to manage the queue. For the requested call in the queue, it searches the database of the remote function entries to schedule the requests to the remote nodes. When the results are sent back, the scheduler thread receives the results and stores it into the output argument of the call. Then, it resumes the waiting thread. The current implementation uses a simple round-robin scheduling. The machine file contains the maximum number of jobs as well as the list of host names. When all remote nodes are busy and the number of jobs reaches to the limit, the thread executing OmniRPC_Call() is blocked. As soon as the jobs for requested remote call is over, the next request is scheduled if any waiting requests exist in the queue.

3   OpenMP Client Using OmniRPC

Since OmniRPC is thread-safe, multiple remote procedure calls can be outstanding simultaneously from multi-threaded programs written in OpenMP.

A typical application of OmniRPC in OpenMP is to execute the same procedure over different input arguments as follows:

```
OmniRPC_init(); /* initialize RPC */
....
#pragma omp parallel for
for(i = 0; i < N; i++)
  OmniRPC_Call("work",i,...);
```

In this loop, the remote function work are executed in parallel with different arguments i in the remote nodes.

The procedure-level task-parallelism is also described as in the following code:

```
#pragma omp parallel sections
{
#pragma omp section
  OmniRPC_Call("subA");
#pragma omp section
  OmniRPC_Call("subB");
#pragma omp section
  OmniRPC_Call("subC");
}
```

The OpenMP clients of OmniRPC can be executed in a single processor. In our Omni OpenMP compiler, we need to execute OpenMP program containing OmniRPC calls in the following environment:

- Set the environment variable OMP_SCHEDULE to "static,1", meaning cyclic scheduling with chunk size 1. In the compiler, the default loop scheduling is block-scheduling which may cause load imbalance when the execution time of each remote call changes.
- Set the environment variable OMP_NUM_THREADS to the number greater than the total number of jobs in available remote nodes. The large numbers of threads are needed to issue the remote procedure call simultaneously for many remote nodes. Furthermore, multiple requests to the remote nodes may hide the latency of communication and the overhead of executable invocation by the local scheduler in remote nodes.
- Compile with "mutex-lock" configuration. As default, the Omni OpenMP compiler uses the spin-lock for fast synchronization in a multiprocessor. It, however, sometimes delays the context-switch between threads in a single processor. The mutex-lock configuration uses the mutex lock of the POSIX thread library for better operating system scheduling.

In the recent release OpenMP 2.0, there is a new clause, NUM_THREADS to parallel region directives. This clause requests that a specific number of threads are used in the regions. This also works for nested regions with task-parallelism and loop-parallelism as in the following code:

```
#pragma omp parallel sections num_threads(3)
{
#pragma section
#pragma omp parallel for num_threads(10)
   for(i = 0; i < N; i++) OmniRPC_Call("workA",i);
#pragma section
#pragma omp parallel for num_threads(20)
   for(j = 0; j < N; j++) OmniRPC_Call("workB",j);
#pragma section
   workC();
}
```

4   Current Status and Future Work

In this paper, we proposed a parallel programming model using the thread-safe OmniRPC in an OpenMP client program for a cluster and global computing. The programmer can build a global computing system by using the remote libraries as its components, without being aware of complexities and hassles of network programming. OpenMP provides an ease-to-use and simple programming environment for a multi-threaded client of OmniRPC. Currently, we have finished a preliminary implementation of OmniRPC. We are doing several experiments and evaluations on some parametric search applications.

The current implementation employs a simple round-robin scheduling over available remote nodes. In the grid environment, the computation time of RPCs may be greatly influenced by many factors including computational ability of the nodes, the distance to the nodes with respect to the bandwidth of communication, and the status of the nodes. More sophisticated scheduling using such dynamic information reported by the local job scheduler in the remote node will be required for efficient remote execution.

We are also developing a remote executable management tool for OmniRPC, which sends the IDL of remote functions to generate stubs in remote nodes automatically. It allows the user to install remote libraries without complex and time-consuming install procedure of remote executables for many remote nodes.

References

1. I. Foster and C. Kesselman. Globus: A metacomputing infrastructure toolkit. *International Journal of Supercomputer Applications*, vol.11, No.2, pages 115–128, 1997. http://www.globus.org/.
2. M. Sato, H. Nakada S. Sekiguchi, , S. Matsuoka, U. Nagashima, and H. Takagi. Ninf: A Network based Information Library for Global World-Wide Computing Infrastructure. *Proc. of HPCN'97 (LNCS 1225)*, pages 491–502, 1997. http://ninf.etl.go.jp/.
3. H. Casanova and J. Dongarra. Netsolve: A network server for solving computational science problems. Technical report, University of Tennessee, 1996.
4. http://pdplab.trc.rwcp.or.jp/Omni/

Performance Oriented Programming for NUMA Architechtures*

Barbara Chapman, Amit Patil, and Achal Prabhakar

Department of Computer Science, University of Houston, Houston, TX
{chapman,amit,achal}@cs.uh.edu

Abstract. OpenMP is emerging as a viable high-level programming model for shared memory parallel systems. Although it has also been implemented on ccNUMA architectures, it is hard to obtain high performance on such systems, particularly when large numbers of threads are involved. Moreover, it is applicable to NUMA machines only if a software DSM system is present. In this paper, we discuss various ways in which OpenMP may be used on ccNUMA and NUMA architectures, and evaluate several programming styles on the SGI Origin 2000, and on TreadMarks, a Software Distributed Shared Memory System from Rice University. These results have encouraged us to begin work on a compiler that accepts an extended OpenMP and translates such code to an equivalent version that provides superior performance on both of these platforms.

Keywords: shared memory parallelism, parallel programming models, OpenMP, ccNUMA Architectures, restructuring, data locality, data distribution, Software Distributed Shared Memory

1 Introduction

Among the various programming models available for parallel programming, the shared memory programming model provides ease of programming, as the programmer is freed from the intricacies of communication and data collection. It is also much easier for a compiler to generate parallel code for a shared memory machine from a sequential program with programmer directives than for a distributed memory platform.

However, pure SMPs do not scale beyond 8 or 16 processors unless a prohibitively expensive crossbar is deployed. When more processors are added, the shared memory bus architecture of the overwhelming majority of commercial systems presents a serious bandwidth bottleneck. In response, vendors such as SGI, Compaq, Sun and HP have built systems partitioned into smaller modules. Each such

---

* This work was partially supported by NASA Ames Research Center under contract number ****, and by NSF under grant number NSF ACI 99-82160. Initial experiments were performed while the authors were in residence at ICASE, NASA Langley Research Center. These sources of support are gratefully acknowledged.

R. Eigenmann and M.J. Voss (Eds.): WOMPAT 2001, LNCS 2104, pp. 137–154, 2001.

module is a pure SMP with a high speed interconnect linking it to all other modules, either directly or indirectly. The resulting system, potentially comprised of a large number of modules, is characterized by a memory hierarchy with non-uniform memory access times (NUMA). Systems which provide cache coherency are called cc-NUMA. Commercial examples include: SGI's Origin 2000, Compaq's AlphaServer GS80, GS106 and GS320. On such platforms, processes may directly access data in memory across the entire machine via load and store operations.

On the other hand, clusters of uni-processor or multi-processor machines have physically distinct memories with no hardware support for coherency. It is possible to consider such clusters to be shared memory machines, if we implement a layer of software on them that manages memory consistency across the physically distributed memories [1,12,2,13]. The characteristics of distributed memory systems with a Software Distributed Shared Memory (SDSM) layer on top are similar to cc-NUMA machines.

In our work, we consider how to write scalable OpenMP applications despite the non-uniformity of memory accesses. We show a programming style in which the locality of data and work is taken into account. Unfortunately, this programming style involves making many changes to a program's code, and thus it reduces the benefits of OpenMP with respect to ease of programming. We believe that it is possible to translate a standard OpenMP program with a few extensions into a higher-performing equivalent code for ccNUMA platforms. The paper outlines the language extensions needed to support this task. The paper is organized as follows: we first describe ccNUMA architectures and then discuss OpenMP language extensions and strategies provided by the vendors to support efficient execution on ccNUMA systems. We then introduce our experiments and show the superiority of the coding style that we plan to generate from a simple set of such language extensions.

2   ccNUMA Architectures

A typical ccNUMA platform is made up of a collection of Shared Memory Parallel (SMP) nodes, or modules, each of which has internal local shared memory; the individual memories together comprise the global memory. The entire memory is globally addressed, and thus accessible to all processors; however non-local memory accesses are more expensive than local ones. The memory hierarchy thus consists of one or more levels of cache associated with an individual processor, a node-local memory, and remote memory, main memory that is not physically located on the accessing node. A cache-coherent system assumes responsibility not only for fetching and storing remote data, but also for ensuring consistency among the copies of a data item. If data saved in a cache is updated by another processor, then the value in cache must be invalidated. Thus such systems behave as shared memory machines with respect to their cache management schemes.

Our experiments have been performed on the Silicon Graphics' Origin 2000, a representative of such systems [9]. It is organized as a hypercube, where each node typically consists of a pair of MIPS R12000 processors, connected through a hub, together with a portion of the shared memory. Multiple nodes are connected in a hypercube through a switch-based interconnect. One router connects up to 8 processors, since two pairs are connected directly via their hubs; two routers are needed to connect 16 processors, 4 for 32 processors and so on. Each MIPS R12000 processor has two levels of two-way set associative cache, where the first level of cache is on-chip and provides 32KB data cache and 32KB instruction cache (32-byte line size). The second level of cache is off-chip; it typically provides 1-4MB unified cache for both data and instructions (128-byte line size). All caches utilize a Least Recently Used (LRU) algorithm for cache line replacement. In addition, each node contains up to 4GB of main memory and its corresponding directory memory and has a connection to a portion of the I/O subsystem.

Page migration hardware moves data into memory close to a processor that frequently accesses it, thus increasing the data locality. The hub maintains cache coherence across processors using a directory-based invalidation protocol. While data only exists in either local or remote memory, copies of the data can exist in various processor caches. Keeping these copies consistent is the responsibility of the cache-coherent protocol of the hubs. The directory-based coherence removes the broadcast bottleneck that prevents scalability of the snoopy bus-based coherence. Latency of access to level 1 cache is approximately 5.5ns; for level 2 cache the latency is 10 times this amount. Latency of access to local memory is another 6 times as expensive, whereas latency to remote memory ranges from up to twice that for local memory, when at most 1 router is involved, to nearly 4 times the cost of access to local memory when 16 routers are configured. SGI reports that the bidirectional bandwidth is ca. 620 MBps for up to 3 routers (32 processors) and thereafter is ca. 310 MBps. However, the experienced cost of a remote memory access depends not only on the distance of its location, i.e. the number of hops required, but also on contention for bandwidth. Contention can have a severe impact on performance; it can arise as the result of many non-local references within a single code, or may be caused by the activities of other independent applications running on the same machine, and its effect is thus unpredictable.

The operating system supports data allocation at the granularity of a physical page. It attempts to allocate memory for a process on the same node on which it runs. However, results are not guaranteed. Default strategies may be set by the user or the site. Typically, a default first-touch page allocation policy is used that aims to allocate a page from the local memory of the processor incurring the page-fault. In an optional round-robin data allocation policy, pages are allocated to each processor in a round-robin fashion.

2.1   TreadMarks: Software Distributed Shared Memory System

TreadMarks [1,12] is a Software Distributed Shared Memory system (SDSM) that uses the operating system's virtual memory interface to implement the shared memory abstraction. It works on most UNIX systems that provide a mechanism for a user process to get memory violation notifications. To maintain memory consistency, TreadMarks employs an extension to the Release Consistency protocol (RC) [3] called the Lazy Release Consistency protocol [8].

RC is a relaxed memory consistency model that allows a processor to delay making its changes to the shared data visible to other processors until a certain synchronization point is reached. This reduces the overall number of messages that must be transferred and allows the messages to be grouped together with data transfer. Lazy Release Consistency (LRC) [8] is an extension of RC wherein the propagation of modifications is postponed until the time of the acquire. A release in LRC is a completely local event and requires no communication. However, at an acquire, the acquiring processor must determine which modifications it needs from which processors, according to the definition of RC. The lazy implementation aims at reducing the number of messages and the amount of data transferred.

To eliminate the effects of false sharing, TreadMarks uses the multiple-writer protocol. With this protocol, two or more accesses can simultaneously occur on the same page and the modifications are done on a local copy of the shared page. When the processors reach a synchronization point, they exchange the diffs created by comparing the original copy to the modified local copy. The processors then apply each other's diffs on the local copy. In TreadMarks, the diffs are created only when requested by a processor, and not at every release and acquire. This lazy diff creation helps reduce the overall number of diffs created and can improve the performance.

The behavior of a distributed memory platform programmed via the TreadMarks system is similar in spirit to that of ccNUMA systems, and any technique that is expected to give better performance on a ccNUMA machine, is also expected to yield some performance gain for a SDSM system.

3   OpenMP

OpenMP consists of a set of compiler directives, as well as library routines, for explicit shared memory parallel programming. The directives and routines may be inserted into Fortran, C or C++ code in order to specify how the program's computations are to be distributed among the executing threads at run time. It provides a familiar programming model, enables relatively fast, incremental and portable application development, and has thus rapidly gained acceptance by users. The OpenMP directives may be used to declare parallel regions, to specify the sharing of work among threads, and for synchronizing threads. Worksharing directives spread loop iterations among threads, or divide the work into a set of parallel sections. Thus it is easy to specify task parallelism. Parallel regions may differ in the number of threads assigned to them at run time, and the assignation of work to threads may also be dynamically determined. It is thus relatively easy to adapt an OpenMP program to a fluctuating computational load, or even to a changing workload on the target platform. Users may set the number of executing threads; typically, there will be one thread per executing processor at run time.

3.1   OpenMP Language Extensions for ccNUMA Platforms

Although OpenMP can be transparently implemented on a ccNUMA platform, as well as mapped to a SDSM system such as TreadMarks, it does not account for non-uniformity of memory access by design. Therefore, the user cannot explicitly specify that data should be allocated on or near a node where computations based upon it are performed; nor can the user explicitly prefetch data during execution within OpenMP.

The best solution to the problem of co-allocating data and threads in an OpenMP program would be to implement a transparent, and highly optimized, dynamic migration of data. However, it is very hard for the operating system to determine when to migrate data and current commercial implementations do not perform particularly well. Moreover, page-based storage is not always a suitable basis for an appropriate distribution of data. Both SGI and Compaq thus provide low-level features for directly influencing the location of pages in memory, as well as high level directives to specify data distribution and thread scheduling in OpenMP programs [4,7]. The extension sets differ. It is unfortunate that their syntax also differs, since there is substantial overlap in the core functionality of the two sets of directives. A major component of both sets is the DISTRIBUTE directive. This specifies the manner in which a data object is mapped onto the system memories. Three distribution kinds, namely BLOCK, CYCLIC and *, are available to specify the distribution required for each dimension of an array. For example, in SGI FORTRAN the following statement will distribute the two dimensional array A by block in the first dimension:

!$SGI DISTRIBUTE A(BLOCK,*)

The Compaq directive corresponding to the above directive is:

!DEC$ DISTRIBUTE A(BLOCK,*)

These directives influence the virtual memory page mapping of the data object, hence the granularity of distribution is limited by the granularity of the underlying pages, which is at least 16KB on the SGI Origin 2000. The advantage is that these directives can be added to an existing program without any restrictions, since they do not change the arrangement of the data object itself. A major disadvantage is that they are unsuitable for distributing small arrays. Both sets of extensions also provide a REDISTRIBUTE directive, with which an array distribution can be changed dynamically at run-time.

It is possible to perform data distribution at element granularity rather than page granularity. This involves rearranging the layout of the array in memory so that two elements which should be placed in different processor's memories are stored in separate pages, which may require padding the array or other arrange-ments. The resulting data layout may violate the standard language array layout specifications, but it guarantees the specified distribution at the element level.

The data mappings are defined as in the HPF standard. The SGI FORTRAN directive for this is:

!$SGI DISTRIBUTE_RESHAPE A(BLOCK,*)

Compaq requires that the NOSEQUENCE directive be supplied along with the DISTRIBUTE directive in order to specify element granularity. There are some limitations on the use of elementwise distributed arrays [7], as a result of the non-standard layout in memory.

Both vendors also supply directives to associate computations with the location of data in storage. Compaq provides the NUMA directive to indicate that the iterations of the immediately following PARALLEL DO loop are to be assigned to threads in a NUMA-aware manner, i.e. according to the distribution of the data. It may distribute multiple levels of the loop, in contrast to the single level specified by the OpenMP standard. The ON HOME directive informs the compiler exactly how to distribute iterations over memories, ALIGN is used for specifying alignment of data, MEMORIES is roughly equivalent to the HPF PROCESSORS directive, although it maps data to local memories rather than individual processors, and the TEMPLATE directive is used to define a virtual array. SGI similarly provides AFFINITY, a directive that can be used to specify the distribution of loop iterations based on either DATA or THREAD affinity (thus closely related to the ON HOME directive of Compaq) and the NEST directive, corresponding to Compaq's NUMA.

We have begun to implement a set of directives that is similar to the above; indeed, we have used them as the starting point for our work. We aim to use directives that are as simple as possible without sacrificing important functionality. Our set therefore includes the above data distribution features, but also includes a general block distribution that may help map the data of some unstructured codes.

The use of a PROCESSORS directive enables load-balanced mappings to be specified for systems where the nodes are heterogeneous, but it does not capture the structure of a hierarchical system. In contrast, the MEMORIES directive enables one to identify and target the individual SMPs in a large system, but causes problems if the individual nodes have different numbers of configured processors. A general block distribution can help achieve load balance in the latter case, so would seem to provide a way out of this apparent dilemma. However, our compilation strategy considers data to be ultimately mapped to threads, since it can then be privatized.

An ON HOME directive similar to that of Compaq maps iterations or parallel sections to a processor associated with the memory a particular datum is stored on, and thus augments OpenMP's own mechanisms for distributing the computation among threads. If no mapping is explicitly specified, OpenMP defaults prevail. Further, our set of directives includes the SHADOW directive borrowed from HPF. This directive enables the user to specify the extent of non-local data accessed by a thread, whether it is read or written, during the computation assigned to it. It will be used by the compiler to set up buffers for storing and copying this data.

Our goal is to translate these directives to an SPMD-style of program. The approach was motivated not only by a variety of previous reports on OpenMP programming experiences, but also by experiments we have conducted using OpenMP on an Origin and on top of TreadMarks. We describe these experiments below. Although the set of programs is small, the examples were chosen to reflect different kinds of data access patterns. On-going work investigates the translation of more substantial codes. Implementation of an initial set of directives has begun in the Cougar compiler under development at the University of Houston.

4   Standard OpenMP Programming

We describe experiments using three sample applications written in OpenMP in a straightforward loop-level parallel programming style. For our ccNUMA platform, we relied on the Origin 2000 systems at the National Center for Supercomputing Applications (NCSA). They were compiled with SGI's MIPSpro7 Fortran 90 compiler under the options -mp -64 -mips4 -r1000 -Ofast -IPA and run in multiuser mode. We made use of the various strategies provided by SGI to obtain good performance under OpenMP. Initially, the most suitable "first touch" default strategies, whereby a datum is stored on the node where it is first accessed, were used in each case. In all but one case, the results shown were obtained with the Origin's _DSM_MIGRATION (page migration) environment variable was set to OFF. One set of results with this switch enabled is provided, so that the reader may compare them.

The SDSM experiments make use of the SDSM system TreadMarks, which was installed on an IBM SP2 system at the University of Houston; it operated on thin nodes with 128MB memory and running at 120MHz. The code was compiled with the IBM C compiler version 3.6.6 and linked with Tread-Marks runtime library verion 1.0.3.3-BETA. The compiler options used were -O3 -qstrict -qarch=pwr2 -qtune=pwr2.
This platform has a relatively slow interconnect; therefore it is not a surprise that the speedups obtained are generally low, and in particular, much lower than those on the SGI Origin. But the gains in speedup that we shall see as a result of a distribution and privatization of data are greater in this case, since there is a much higher difference in cost between local and remote memory access.

The speedup figures shown in this and the following section have all been normalized. They are speedups with respect to the serial time of the initial OpenMP version (without vendor-specific directives or multiprocessing code option on the O2000). We have not included results for more than 64 (Jacobi) and 32 (LU, LBE) processors respectively, since for the matrix sizes used, there was not enough computation remaining to keep additional threads busy.

4.1   Jacobi

The Jacobi method is one of the simplest numerical solvers for partial differential equations. Although it converges very slowly, it exhibits excellent spatial locality. In this short code, almost all data accesses can be made local (see Prog. 1 and Fig. 4.1 (a)). Thus it should be possible for a system to obtain good performance on a Jacobi stencil under OpenMP without user-specified distribution directives. On the Origin, we may exploit the default first touch page allocation policy to ensure that pages are distributed among the executing processors. There will be shared read access to the boundary columns. The use of SGI's DISTRIBUTE (*,BLOCK) directive will do exactly the same thing, the only difference being that it can be set up at compile time. We give timings for the default allocation as well as for columnwise BLOCK distribution of matrices A and B using SGI's DISTRIBUTE and DISTRIBUTE_RESHAPE directives.

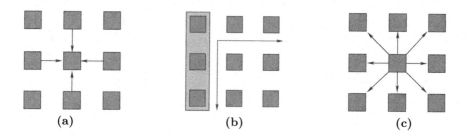

(a)   **(b)**   **(c)**

Fig. 1. Jacobi (a), LU (b) and LBE (c) access stencils

```
!$OMP PARALLEL DO
  do j = 1,n
    do i = 1,n
      A(i,j) = (B(i-1,j)+B(i+1,j)+B(i,j-1)+B(i,j+1)) * c
    end do
  end do
!$OMP END PARALLEL DO
```

Program 1. Standard OpenMP version of Jacobi Kernel

The TreadMarks version of the Jacobi kernel is a straightforward translation of the corresponding OpenMP version. Shared variables are allocated explicitly using the Tmk_malloc primitive. Iteration space is block divided and synchronization is achieved through calls to the Tmk_barrier function.

The performance figures shown are based on runs with a 1024 by 1024 matrix of double precision data. For this data size, the local portion of the matrix fits precisely into a number of pages for all the numbers of threads used. Note, however, that only the elementwise DISTRIBUTE_RESHAPE directive guarantees that data will start on a page boundary. No matrix element is updated by more than one thread. All but the first and last threads must read 2 columns that are updated by another thread. If we ensure that the updating thread is the first to reference the data, the first touch policy will realize a pagewise block distribution of the second dimension. Threads will need to read 2 * 1024 elements (two half-pages, or 128 cache lines) that are stored in proximity to another thread at each iteration.

We first show results obtained with the _DSM_MIGRATION switch set on (Fig. 4.1 (a)). If we compare this with our other results (Fig. 4.1 (b)), it is clear that pages are indeed being migrated; they will have to migrate back to their previous location for updating. With an increasing number of processors, more and more pages are moved; from 16 processors onward, some data will move across one or more routers.

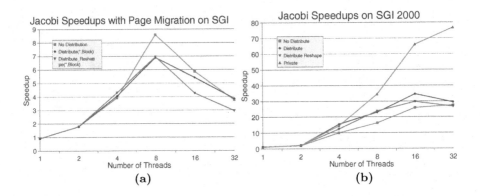

Fig. 2. OpenMP Jacobi speedups on SGI O2000 with page migration (a) without page migration (b)

The first touch policy (Prog. 1) and DISTRIBUTE (*,BLOCK) realize the same data mapping, and since, in this example, each array element will be updated by only one processor, the page mapping remains the same after initialization in both cases. The compile time mapping enabled by the user directive provides better performance when the number of processors is increased. Note that we were only able to achieve the benefit of the DISTRIBUTE_RESHAPE (*,BLOCK) directive when using the *fast* compiler switch. Without it, performance was poor since the compiler did not optimize the pointer arithmetic introduced to realize accesses to the array elements in this version. Performance figures are consistently good for all three versions of the Jacobi code on the Origin.

The speedups obtained with Treadmarks, shown as "shared" code in Fig. 4.1 are not impressive. In fact, there is a slowdown for more than 16 processors. This is attributed mainly to the high coherency maintenance overheads. Many

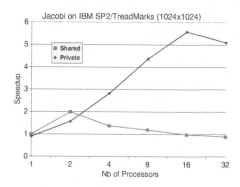

Fig. 3. Jacobi speedups on IBM SP2 with TreadMarks

pages are shared and TreadMarks has to collect and disseminate coherency data for all of them at each synchronization point. We discuss the "private" version further below.

4.2   LU

Our second experiment performed an LU factorization, another standard numerical solution method, in which a matrix is factorized into upper and lower triangular matrices. As with the Jacobi code, our program only requires shared read access, and that only on one column per iteration. However, the column involved differs in each iteration. The simple version used here does not implement pivoting (see Prog. 2).

The matrix size operated on reduces progressively (see Fig. 4.1 (b)). Therefore, we ensure that all threads continue to participate in the computation and ensure good load balancing by using a CYCLIC distribution. This can be achieved using the page-based DISTRIBUTE directive or the elementwise DISTRIBUTE_RESHAPE together with the CYCLIC data distribution in the second dimension. The AFFINITY clause is needed in order to ensure that the DO loop is distributed according to the data mapping. Otherwise, the default mapping would assign iterations to threads by block. Timings for the code without the AFFINITY directive were approximately 40 % higher than those shown here.

In the TreadMarks version, compute matrix lu is declared shared and iterations are block distributed. The normalization code is protected by Tmk_lock_acquire and Tmk_lock_release primitives.

We show results for our LU computation based upon runs with a 2048 by 2048 real matrix in Fig. 4.2. For the version without distribution directives, the speedup is less than linear after 16 processors are used. The other two versions continue to have nearly linear speedup on 32 processors. In the former case, the initial data mapping will lead to an increasing load imbalance, as well as to larger numbers of remote fetches as the computation progresses. The DISTRIBUTE and DISTRIBUTE_RESHAPE versions perform somewhat better, since the distribution

```
!$OMP PARALLEL
   do k = 1,n-1
!$OMP SINGLE
       lu(k+1:n,k) = lu(k+1:n,k)/lu(k,k)
!$OMP END SINGLE
!$OMP DO
       do j = k+1, n
           lu(k+1:n,j) = lu(k+1:n,j) - lu(k,j) * lu(k+1:n,k)
       end do
!$OMP END DO
   end do
!$OMP END PARALLEL
```

Program 2. Standard OpenMP versions of LU

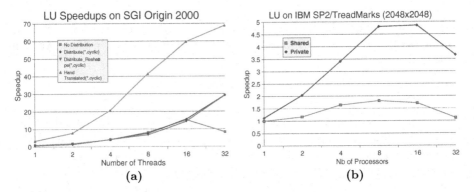

Fig. 4. LU speedups (a) on SGI O2000 (b) IBM SP2 with TreadMarks

reflects the load-balancing requirements.

The inherent load imbalance in the LU algorithm leads to contention at the node that performs the normalization. Coupled with the high coherence overheads due to the large number of shared pages, this lead to the performance degradation in the TreadMarks version.

4.3   LBE

Our third experiment makes use of LBE, a computational fluid dynamics code that solves the Lattice Boltzmann equation. It was provided by Li-Shi Luo of ICASE, NASA Langley Research Center [16]. The numerical solver employed by this code uses a 9-point stencil. However, unlike the Jacobi solver, the neighboring elements are updated at each iteration (cf. Fig. 4.1 (c)). The Collision_advection_interior subroutine with a write shared data access is shown in Program 3. The shared memory policy permits multiple reads on the same cache line (or page) but not multiple writes. Thus LBE allows us to analyze this weakness of ccNUMA architectures. As before, we developed three versions of this code. In the versions with distribution directives, the matrices are BLOCK distributed in the last dimension to ensure good data locality and an even load balance. The TreadMarks version for the LBE solver is similar in spirit to the Jacobi code.

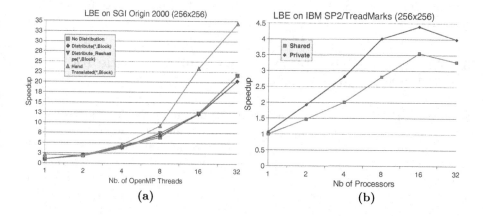

Fig. 5. LBE speedups (a) on SGI O2000 (b) on IBM SP2 with TreadMarks

```
                                   Collision_advection_interior:
!$OMP PARALLEL                       !$OMP DO
    do iter = 1, niters                 do j = 2, Ygrid-1
        Calculate_ux_uy_p                   do i = 2, Xgrid-1
        Collision_advection_interior            f(i,0,j)   = Fn(fold(i,0,j))
        Collision_advection_boundary            f(i+1,1,j) = Fn(fold(i,1,j))
    end do                                      f(i,2,j+1) = Fn(fold(i,2,j))
!$OMP END PARALLEL                              f(i-1,3,j) = Fn(fold(i,3,j))

                                                . . . . .

                                            f(i+1,8,j-1) = Fn(fold(i,8,j)
                                        end do
                                    end do
                                 !$OMP END DO
```

Program 3. OpenMP version of the LBE Algorithm and Collision_advection_interior Kernel

The experiments were performed with a 256 by 256 matrix (see Fig. 4.3). On the Origin, all three versions behaved similarly up to 32 processors. Speedups are lower than those previously reported, a result of the fact that each processor has to access the (logical) neighbor's memory for writing rather than reading as in the case of Jacobi or LU. Cache lines at the boundaries of local data will ping-pong between a pair of processors where the corresponding threads execute. For 16 to 32 processors, remote data fetches might involve 2 hops through the network, and thereby longer remote access times may be observed.

The speedups obtained for the TreadMarks version of LBE are significantly better than Jacobi and LU. They increase consistently till 16 processors, beyond which the per processor computation is not enough to offset the communication overhead.

5   An Alternative OpenMP Programming Style for ccNUMA Platforms

Under the SPMD parallelization strategy, we distribute the arrays among the processors and convert the local part into an array that is private to each thread. One or more shared buffers are created to exchange data as needed between the threads. For instance, if one processor must read a row that is stored in the private memory of the "neighbor" thread, a shared array with the size of a row is created as a buffer to copy in this data. Once data has been copied into the shared buffer, it may be written into an array that is private to the processor that needs it. This is most efficiently performed by extending the size of each private array to include the "shadow" regions, as is also realized via a SHADOW directive in HPF. The programmer must explicitly synchronize reading and writing of buffer data. Thus each thread works on its private data, and sharing is enabled through small shared buffers. The resulting code resembles an MPI program to some extent, but it is easier to specify and potentially provides better performance than the corresponding MPI code [15].

The SPMD style has been used in conjunction with OpenMP by other researchers, but more often codes have used a mixture of OpenMP and MPI for better portability (cf. [10,14,6,11]). In [15], a comparison of SPMD OpenMP, MPI and Co-Array Fortran programming styles is based on an ocean model. The author points out that using halos (or shadows) and an SPMD style also reduces false sharing.

In the corresponding TreadMarks code versions, labeled *private* in the performance figures shown above, the arrays are privatized to realize this SPMD style (cf. the hand-coded versions presented below). Codes are y similar to the corresponding Origin versions (e.g. LU code in Prog. 6for TreadMarks and in Prog. 5 for SPMD OpenMP).The results we obtained are given in Figures 4.1 for Jacobi, 4.2 (b) for LU and 4.3 (b) for LBE.

Jacobi in SPMD style: In order to create this version of our Jacobi code, we divide arrays A and B among the processors in the second dimension, and create pairs of buffers for exchanging data at the two boundaries (one each for the first and last column). The size of the second dimension of the private arrays A and B is now ($n + 2$ /no. of threads), where n is the original size, and space is reserved for the shadow area. The size of the first dimension remains the same as in the shared version. The shared buffers are of size $(2*N-2)*$ *columnsize*, where N is the number of threads. At the start of the PARALLEL region the buffers are declared to be shared, and A and B are declared to be private to each thread.

First, a thread copies its first and last column into the appropriate columns of the shared buffers to initialize them. This is followed by an OpenMP BARRIER so that no thread accesses the buffers until they have been written. The threads then copy appropriate columns from the buffers into their shadow area. The rest of the Jacobi kernel remains the same as before, with each thread executing the code to calculate its own portion of data.

```
!$OMP PARALLEL SHARED(buflower,bufupper,chunk) &
!$OMP PRIVATE(A,B,thdno,noofthds,start_x,end_x,start_y,end_y,i,j,k)
      .......
    do k = 1, ITER
      bufupper(1:n,thdno) = A(1:n,chunk)
      buflower(1:n,thdno) = B(1:n,1)
!$OMP BARRIER
      B(1:n,0) = bufupper(1:n,thdno-1)
      B(1:n,chunk+1) = buflower(1:n,thdno+1)
      do j = start_y, end_y
        do i = start_x, end_x
          A(i, j) = (B(i-1,j) + B(i+1,j) + B(i,j-1) + B(i,j+1)) * c
        enddo
      enddo
      do j = start_y, end_y
        B(1:n, j) = A(1:n, j)
      enddo
    enddo
!$OMP END PARALLEL
```

Program 4. Jacobi Kernel in SPMD style

This version of the program shows superlinear speedup (Fig. 4.1 (a)). Not only is local cache used efficiently, there is less intrusion from the operating system while handling shared variables. Shared buffers are updated only once per iteration, and the update occurs separately from the rest of the computation. The SGI compiler is able to do a particularly good job of optimizing such "vector" updates, so that data transfer cost is further reduced.This good performance continues up to 128 processors, although speedup is no longer linear, as a result of the decreasing computation per thread.

The TreadMarks results for the private version show speedups which are much better than the shared version. Increased data locality and lower coherency overheads improve the computaion to communication ratio i.e more percentage of the execution time is spent doing useful computation. For the private case, we get an almost linear speedup upto 16 processors. With 32 processors there is a decrease in speedup.

LU in SPMD style: This code uses only one shared buffer of the size of one column to share the pivot column among the processors. Each thread requires a private buffer of the same size. We realize the cyclic data distribution previously chosen by assigning columns to threads in a round robin fashion. The size of the second dimension will be *n / No. of threads*), where *n* is the original size of this dimension.

In the PARALLEL region (see Prog. 5), each column is chosen in turn and normalized. After normalizing, the pivot column is copied to the shared buffer. All other threads wait at the barrier. They then copy the contents of the shared buffer to their private buffer. The rest of the computation in the kernel involves only the columns to the right of the normalized one, and each thread can perform its portion independently on its own data.

```
!$OMP PARALLEL SHARED(shbuf)&
!$OMP PRIVATE(lu,sumP,counter,noofthds,id,i,j,K,pribuf,Npri)
   DO K=1,N-1
! Compute Column K
       if(mod(K-1,noofthds).eq.id) then
           lu(K+1:N,counter)  = lu(K+1:N,counter) / lu(K,counter)

!move the column K to the sharedbuffer
           shbuf(K+1:N) = lu(K+1:N,counter)
           counter = counter+1
       end if
!$OMP BARRIER
!move the sharedbuffer to private buffer or column 0
       pribuf(K+1:N) = shbuf(K+1:N)
! Update Right Part (Column J+1:N)
       do j = counter, Npri
         lu(K+1:N, j) = lu(K+1:N, j) - lu(K, j) * pribuf(K+1:N)
       end do
   END DO
!$OMP END PARALLEL
```

Program 5. LU Kernel in SPMD style

The only sequential part of this version is when the pivot column is copied by the thread that "owns" it to the shared buffer, while the other threads wait at the barrier. The rest of the computation is completely parallel, with each thread working on its private array. The super linear speedup (Fig. 4.2 (a)) can be explained by the cache effect and efficient handling of the shared buffer updates.

For the TreadMarks version only one column was declared shared, the rest of the array was divided cyclicly and made totally private. The iteration space was also divided in cyclic fashion. This version is consistently better than the shared version.

```
for(ck=0,k=Tmk_proc_id; ck < N-1 ;ck++)
{
  if(ck == k)
  {
    for(j=ck+1;j<N;j++)
      shared->lu[j] = lu[ck/Tmk_nprocs][j] =
        lu[ck/Tmk_nprocs][j] / lu[ck/Tmk_nprocs][ck];
      k = k + Tmk_nprocs;
  }
  Tmk_barrier(1);
  /* update right part. */
  for(j=k;j<N;j+=Tmk_nprocs)
    for(i=ck+1;i<N;i++)
      lu[j/Tmk_nprocs][i] = lu[j/Tmk_nprocs][i] -
                            lu[j/Tmk_nprocs][ck]*shared->lu[i];
  Tmk_barrier(2);
}
```

Program 6. LU Factorization with TreadMarks

LBE in SPMD style: The SPMD version of LBE realizes a BLOCK distribution of the arrays ux, uy, p, f, fold. Only the array f, which has a shared write access, requires a shadow column on either side of the private data. In contrast to the previous codes, data is written by a neighboring thread and must be subsequently read by its "owner". After the shadow columns are updated in subroutine Collision_advection_boundary, their contents are copied to shared buffers. After the copy has completed, the owner of the data may copy it into its private array. The synchronization needed to do this is the only difficult part when creating the SPMD code.

The SPMD version of LBE shows a remarkable increase in performance for 16 and 32 processors (Fig. 4.3 (a)). In this version, the processors work in parallel on their private data till they reach the end of the iteration. Previously, non-local updates, and the copying of cache lines at boundaries, occurred throughout the computation. Here, sharing is again restricted to the buffer arrays, and it occurs at specific points, in a manner that easily permits compiler optimization. The speedup is lower beyond 32 processors, as a result of the relatively small matrix size used in the experiment (256 by 256).

In all cases, the SPMD program versions exhibit better speedup than those obtained by using the SGI directives DISTRIBUTE or DISTRIBUTE_RESHAPE. The translation from the loop parallel OpenMP code to the SPMD OpenMP mode required the selection of a distribution strategy for the program's data. After computing the local size of data objects, including shadow regions, it is a matter of introducing buffer copying and synchronization to ensure that data is read from buffers after it has been written to them.

A comparison of the different code versions under TreadMarks demonstrates that this programming style can make a significant difference, even for software distributed shared-memory systems. For the Jacobi kernel, the privatized version outperforms the shared version by more than a factor of 5 on 16 threads; for the LU code, it is a factor of 4. The difference is not so large for the LBE kernel, and corresponds to the results on the SGI Origin.

6   Conclusions and Future Work

OpenMP [5] is a set of directives for developing shared memory parallel programs. It is an effective programming model for developing codes that are to run on small shared memory systems. It is also a promising alternative to MPI for codes running on ccNUMA platforms – or it would be, if it could provide similar levels of performance on these together with ease of programming.

We have shown the performance benefits of a programming style that privatizes data in the experiments discussed in this paper. This style relies on the partitioning of global data to create local, private data for each thread, and the corresponding adaptation of loop nests. It also requires the explicit construction of buffer arrays for the transfer of data between threads. It further obtains performance by mapping loop iterations to threads for which specified data is local, in the sense that it is stored in local memory. At the University of Houston, we are beginning to develop a source-to-source compiler for OpenMP that will accept a modest set of extensions to OpenMP and use them to generate an OpenMP code written in this style. The most important of these are the (DISTRIBUTE), (ON HOME), and SHADOW directives.

Acknowledgements. The authors wish to thank Li-Shi Luo for the provision of the LBE code that was by far the most interesting of those studied in this paper, and Piyush Mehrotra and Seth Milder at ICASE, NASA Langley Research Center,for their help in understanding the application and the problems it poses. They are also grateful to Jerry Yan and Michael Frumkin for their encouragement to investigate performance problems related to this architecture, and to Tor Sorevik, who introduced the first author to OpenMP and its intricacies on the Origin 2000.

References

1. C. Amza, A. Cox, and et al. Treadmarks: Shared memory computing on networks of workstations. IEEE Computer, 29(2):18–28, February 1996.
2. J. Bennett, J. Carter, and W. Zwaenepoel. Munin: Shared memory for distributed memory multiprocessors, 1989.
3. J. Bennett, J. Carter, and W. Zwaenepoel. Munin: Distributed shared memory using multiprotocol release consistency, 1991.
4. J. Bircsak, P. Craig, R. Crowell, Z. Cvetanovic, J. Harris, C.A. Nelson, and C.D. Offner. Extending OpenMP for NUMA Machines. In *SC2000, Supercomputing*, Dallas, Texas, USA, November 2000.

5. OpenMP Architecture Review Board. OpenMP Fortran Application Program Interface, Version 2.0, November 2000.
6. F. Cappello and D. Etiemble. MPI versus MPI+OpenMP on IBM SP for the NAS Benchmarks. In *SC2000, Supercomputing*, Dallas, Texas, USA, November 2000.
7. Silicon Graphics Inc. MIPSPro Fortran 90 Commands and Directives Reference Manual. Document number 007-3696-003. Search keyword MIPSPro Fortran 90 on http://techpubs.sgi.com/library/.
8. P. Keleher, A. Cox, and W. Zwaenepoel. Lazy release consistency for software distributed shared memory. In *19th Annual International Symposium on Computer Architecture*, pages 12–21, May 1992.
9. J. Laudon and D. Lenoski. The SGI Origin ccNUMA Highly Scalable Server. SGI Publishe White Paper, March 1997.
10. P. Kloos and F. Mathey and P. Blaise. OpenMP and MPI programming with a CG algorithm. In *EWOMP 2000, European Workshop on OpenMP*, Edimburgh, Scotland, U.K., September 2000.
11. R. Blikberg and T. Sorevik. Early experiences with OpenMP on the Origin 2000. In *Proc. European Cray MPP meeting*, Munich, September 1998.
12. Concurrent Programming with TreadMarks. TreadMarks users Manual. http://www.cs.rice.edu/ willy/papers/doc.ps.gz.
13. M. Sato, H. Harada, and Y. Ishikawa. OpenMP compiler for a Software Distributed Shared Memory System SCASH. In *WOMPAT 2000*, San Diego, July 2000.
14. L.A. Smith and J.M. Bull. Development of Mixed Mode MPI/OpenMP Applications. In *WOMPAT 2000*, San Diego, July 2000.
15. A.J. Wallcraft. SPMD OpenMP vs MPI for Ocean Models. In *First European Workshop on OpenMP - EWOMP'99*, Lund University, Lund, Sweden, 1999.
16. X. He and L.-S. Luo. Theory of the Lattice Boltzmann Method: From the Boltzmann Equation to the Lattice Boltzmann Equation. In *Phys. Rev. Lett. E, 56(6), 6811*, 1997.

Defining and Supporting Pipelined Executions in OpenMP

M. Gonzalez, E. Ayguadé, X. Martorell, and J. Labarta

Computer Architecture Department, Technical University of Catalonia,cr. Jordi
Girona 1-3, Mòdul D6, 08034 - Barcelona, Spain
{marc, eduard, xavim, jesus}@ac.upc.es

Abstract. This paper proposes a set of extensions to the OpenMP programming model to express complex pipelined computations. This is accomplished by defining, in the form of directives, precedence relations among the tasks originated from work–sharing constructs. The proposal is based on the definition of a name space that identifies the work parceled out by these work–sharing constructs. Then the programmer defines the precedence relations using this name space. This relieves the programmer from the burden of defining complex synchronization data structures and the insertion of explicit synchronization actions in the program that make the program difficult to understand and maintain. The paper focuses on the runtime support required to support this feature and the code generated by the NanosCompiler.

1   Introduction

OpenMP [6] has emerged as the standard programming model for shared–memory parallel programming. One of the features available in the current definition of OpenMP is the possibility of expressing multiple–levels of parallelism. When applying multi–level parallel strategies, it is common to face with the necessity of expressing pipelined computations in order to exploit the available parallelism. These computations are characterized by a data dependent flow of computation that implies serialization. In this direction, the specification of generic task graphs as well as complex pipelined structures is not an easy task in the framework of OpenMP. In order to exploit this parallelism, the programmer has to define complex synchronization data structures and use synchronization primitives along the program, sacrificing readability and maintainability.

In this paper we propose an extension to the OpenMP programming model with a set of new directives and clauses to specify generic task graphs and an associated thread mapping. The paper focuses on describing the functionalities required in the runtime supporting the code generated by the compiler.

2   Extensions to OpenMP

In this section we summarize the extensions proposed to support the specification of complex pipelines that include tasks generated from OpenMP work–sharing

R. Eigenmann and M.J. Voss (Eds.): WOMPAT 2001, LNCS 2104, pp. 155–169, 2001.
© Springer-Verlag Berlin Heidelberg 2001

constructs. The extensions are in the framework of nested parallelism and target the pipelined execution at the outer levels. The part of the proposal corresponding to the thread groups definition comes out of previous work that can be found in [1]. An initial definition of the extensions to specify precedences was described in [2].

2.1   Precedence Relations

Next we present an extension to the OpenMP programming model that allows the specification of precedence relations among the threads that participate in the execution of a parallel construct. The proposal is divided in two parts. The first one consists in the definition of a name space for the tasks generated by the OpenMP work–sharing constructs. The second one consists in the definition of precedence relations among those named tasks.

The NAME clause. The NAME clause is used to provide a name to a task that comes out of a work–sharing construct.

```
C$OMP SECTIONS                        C$OMP DO NAME(name_ident)
C$OMP SECTION NAME(name_ident)        ...
...                                   C$OMP END DO
C$OMP END SECTIONS
```

The name_ident identifier is supplied by the programmer and follows the same rules that are used to define variable and constant identifiers.

In a SECTIONS construct, the NAME directive is used to identify each SECTION. In a DO work–sharing construct, the NAME clause only provides a name to the whole loop. The number of tasks associated to a DO work–sharing construct is not determined until the associated do statement is going to be executed. Depending on the number of available threads and the chunk size applied in the loop scheduling, the loop is broken into a different number of parallel tasks. We propose to identify each iteration of the parallelized loop by the identifier supplied in the NAME clause plus the value of the loop induction variable for that iteration. This means that the name space for a parallel loop will be big enough to name each of the iterations of the loop. The programmer simply defines the precedences at the iteration level. These precedences are then translated to task precedences, depending on the SCHEDULE strategy specified to distribute iterations to tasks.

The PRED and SUCC clauses and directives. Once a name space has been created, the programmer is able to specify a precedence relation between two tasks using their names. This is done by the use of the PRED and SUCC clauses or directives:

```
[C$OMP] PRED(task_id[,task_id]*) [IF(exp)]
[C$OMP] SUCC(task_id[,task_id]*) [IF(exp)]
```

PRED is used to list all the tasks names that must complete their execution before executing the one affected by it. The SUCC directive is used to define all those tasks that, at this point, may continue their execution. The IF clause is used like the already existent clause in the OpenMP programming model. Expression exp is evaluated at runtime in order to obtain a boolean value that determines if the associated PRED or SUCC clause applies.

As clauses, PRED and SUCC apply at the beginning and end of a task (because they appear as part of the definition of the work–sharing itself), respectively. The same keywords can also be used as directives, in which case they specify the point in the source program where the precedence relationship has to be fulfilled. Code before a PRED directive can be executed without waiting for the predecessor tasks. Code after a SUCC directive can be executed in parallel with the successor tasks.

The PRED and SUCC constructs always apply inside the nearest work–sharing construct where they appear. Any work–sharing construct affected by a precedence clause or directive has to be named with a NAME clause.

The task_id identifier is used to identify the parallel task affected by a precedence definition or release. Depending on the work–sharing construct where the parallel task was coming out from, the task_id identifier presents three different formats:

```
task_id = (name_ident) | (name_ident,expr) | (name_ident,expr,expr)
```

When the task_id is only composed of a name_ident identifier, the parallel task corresponds to a task coming out from a SECTIONS work–sharing construct. In this case, the name_ident corresponds to an identifier supplied in a NAME clause that annotates a SECTION construct. When the name_ident is followed by one expression, the parallel task corresponds to a chunk of iterations coming from a parallelized loop. The expression must include the loop induction variable and its evaluation must result in an integer value identifying a specific iteration of the loop. The precedence relation is defined with the chunk of iterations containing the iteration indicated by expr. When two expressions are supplied, the task_id syntax allows the programmer to specify a list of predecessor chunks. The first expression indicates the number of parallel tasks that the programmer wants to name. The second expression supplies a pointer reference pointing to a vector containing different values of the iteration space of the loop from where the parallel tasks are coming out. In order to identify the tasks, each value is translated to the loop chunk executing the iteration corresponding to it.

In order to handle nested parallelism, we extend the previous proposal. When the definition of precedences appear in the dynamic extend of a nested parallel region caused by an outer PARALLEL, multiple instances of the same name definition (given by a NAME clause/directive) exist. In order to differentiate them, the name_ident can be extended with an induction variable in case of an outer parallel loop that causes the replication:

```
name_ident[:ind_var]+
```

Therefore, the `task_id` construct might take the following syntax:

```
task_id = (name_ident[:expr]*) |
          (name_ident[:expr]*,expr)
```

The expression that is extending the `name_ident` construct will be evaluated at runtime. Its evaluation must result in an integer value contained in the iteration space of the parallelized loop that cau sed the multiple instances of the same name definition. The precedences apply to the chunk of work that w ill execute the iteration pointed out by the expression.

3   Runtime Support

In this section we describe the support required from the runtime system to efficiently implement the language extensions to specify precedence relations. The runtime system usually offers mechanisms to guarantee exclusive execution (critical regions), ordered executions (ticketing) and global synchronizations (barriers). The proposal in this paper requires explicit point–to–point synchronization mechanisms. The description of the runtime support for multiple levels of parallelism and thread groups is not included in this paper and can be found elsewhere [1]. All these functionalities are included in the NthLib library [5] supporting the code generated by the NanosCompiler [3]. The following subsections describe the most important implementation aspects of the precedences module in the NthLib library. The runtime description is divided in two sections. The first one covers the aspects that have to be considered when deadling with one-level parallel regions. The second one covers the runtime support multilevel parallel regions.

3.1   Single Level Parallelism

Two main aspects have to be considered to provide support to the programming model defined in Section 2. First, the runtime has to provide a mechanism to synchronize two threads according to the precedences introduced by the programmer. Second, there must be a translation mechanism that allows to dynamically identify the thread executing a chunk of iterations in a parallel loop; this means that the runtime has to be able to identify which thread is executing which iteration of the loop.

Thread Synchronization. Our approach is based on the definition of an address space where threads involved in a precedence relation synchronize. For each pair of named tasks related with precedences, a memory location is allocated. Threads executing these tasks will use this memory location to communicate. This memory location is considered as a dependence counter. The definition of a precedence at runtime implies an increment of this counter. The release of a precedence implies a decrement of the same counter. When a thread checks if a precedence has been released, it spins until the counter reaches zero or smaller value. This counter is contained in what we name the precedence descriptor (described later).

Two routines are provided to define/release precedences at runtime: nthf_def_prec and nthf_free_prec, respectively. The main argument for these routines is a precedence descriptor. For each PRED directive/clause, the compiler injects a call to routine nthf_def_prec. This routine increments the counter contained in the precedence descriptor and spins until the counter reaches zero. For each SUCC directive/clause, the compiler injects a call to routine nthf_free_prec. This routine mainly decrements the counter contained in the precedence descriptor.

Figure 1.b shows the code generated by the compiler for the OpenMP fragment shown in Figure 1.a. The compiler injects the code to create the threads that will execute the parallel sections and the invocation of routine nthf_init_prec to allocate and initialize the precedence descriptor. For each SECTION the compiler injects a runtime call to the routine responsible for the thread creation (nthf_create). After the thread creation is performed, the nthf_block routine is invoked. This routine blocks the executing thread until the work that has been generated terminates. Figure 1.b shows the functions that encapsulate the code to be executed by the threads. Notice that function name_A encapsulating the code for named section A performs a call to routine nthf_free_prec in order to decrement the associated counter. Similarly, routine name_B invokes nthf_def_prec in order to increment the counter. The thread invoking this routine will wait until the counter reaches zero. Both routines receive as an argument the precedence descriptor that contains the counter.

When the precedence relation involves threads executing tasks (chunks) of a DO work–sharing construct, the number of counters is determined at runtime. In particular, as many counters as pairs consumer/producer need to be allocated. For instance, the code shown in Figure 2 establishes a precedence between a SECTION and a DO work–sharing construct; in this case, as many counters as the number of threads participating in the loop execution are allocated in the precedence descriptor. These counters are used to synchronize the thread executing the section with each possible thread executing a chunk of the DO loop. In addition to that, the compiler needs to insert code to perform a translation from the iteration space to the thread space. This translation (explained later in this section) determines which thread executes a particular iteration of the loop. The output of the translation determines which counter has to be used for the synchronization.

In the general case, the precedence relation may involve tasks (chunks) generated from two DO work–sharing constructs. In this case, a matrix of counters is allocated (with as many rows and columns as the number of threads) in the precedence descriptor. A routine nthf_init_prec is offered by the runtime to allocate and define the precedence descriptor for each of the above mentioned situations.

```
!$OMP PARALLEL SECTIONS
!$OMP SECTION NAME(A)
        CODE_1
!$OMP SUCC (B)
        CODE_2                      integer*8 prec_A_B(9)
!$OMP SECTION NAME(B)               ...
        CODE_3                      call nthf_init_prec(prec_A_B,...)
!$OMP PRED (A)                      call nthf_create(name_A,prec_A_B,...)
        CODE_4                      call nthf_create(name_B,prec_A_B,...)
!$OMP END PARALLEL SECTIONS         call nthf_block()
        ...
```

a) Source code.   b) Thread creation.


```
subroutine name_A(prec_A_B, ...)        subroutine name_B(prec_A_B,...)
...                 ...
CODE_1                CODE_3
call nthf_free_prec(prec_A_B, ...)  call nthf_def_prec(prec_A_B,...)
CODE_2                CODE_4
...                 ...
end                                     end
```

c) Code executed by each thread.

Fig. 1. Source code and code generated by the compiler for a single level example.

Iteration–Thread Translation. In this section we describe the basic data structure and service available to perform the translation between iteration number and thread executing the iteration. We will focus on the information that is needed at runtime and how this information is used to compute the translation. For each parallelized loop involved in a precedence relation, a loop descriptor is created with the following information: lower and upper bound for the induction variable, iteration step, the scheduling applied and the number of threads currently executing the loop. This loop descriptor is allocated in the application address space. Once a thread participating in the execution of the loop starts executing the loop descriptor is initialized with all the information mentioned above.

The runtime library offers routine nthf_index_to_thread. The main arguments of this routine are a loop descriptor and an iteration index. The output of the routine is the identifier of the thread executing the iteration, which is computed according to the information contained in the loop descriptor.

For example, Figure 2 shows the code generated by the compiler for the example in the same figure. In the code generated for section name_A, the compiler injects a call to routine nthf_index_to_thread in order to know the identifier of the thread executing iteration iter. After that, the thread invokes nthf_free_prec over the corresponding element of the precedence descriptor. For the threads

```
!$OMP PARALLEL                       subroutine name_A(A_loop,loop_desc,...)
!$OMP SECTIONS                       CODE_1
!$OMP SECTION NAME (A)               nth_id = nthf_index_to_thread(loop_desc,
      CODE_1                       1   iter,...)
!$OMP SUCC(loop, iter)               call nthf_free_prec(A_loop,nth_id,...)
      CODE_2                         CODE_2
!$OMP SECTION                        end
      CODE_3                     b) Thread code for the named section
!$OMP END SECTIONS NOWAIT
!$OMP DO NAME (loop)                 subroutine loop(A_loop,iter,...)
      do i = 1, N                    nth_whoami = ...
         CODE_4                      nth_min = ...
!$OMP PRED (A) IF(i.eq.iter)         nth_max = ...
         CODE_5                      do i = nth_min, nth_max
      enddo                             CODE_4
!$OMP END PARALLEL                      if (i.eq.iter) then
                                           call nthf_def_prec(A_loop,
                                     1      nth_whoami,...)
                                        end if
a) Source code                          CODE_5
                                     enddo
                                     end
```

c) Thread code for the loop

Fig. 2. Code generated for a single level example.

executing the loop, notice that the call to routine **nth_def_prec** will be done by the thread executing iteration **iter**. The invocation to **nthf_free_prec** is done with the identifier **nth_whoami** of the thread executing that iteration.

3.2 Multilevel Parallelism

In this section we extend the synchronization requirements described in the previous section to cover the multilevel case. By the use of an example we will point out the runtime requirements to support the proposal described in Section 2. Figure 3.a shows an example of a multilevel code with precedences. Figure 3.b shows the precedence graph at iteration level. Figure 3.c defines the thread identifiers for the example in order to make easier the explanation. Notice that each thread has been identified according to the level of parallelism where it is executing. Threads executing the outer level of parallelism are identified with (0) and (1). Threads executing the inner level are identified with (0,0) (0,1) (1,0) and (1,1). Figure 3.d shows the iteration space defined for each thread, taking into account that OMP_NUM_THREADS=2. Finally, Figure 3.e shows the precedence graph at the thread level.

Consider the thread executing the following portion of the iteration space: iterations 1-2 of the j loop coming from the parallel branch in the outer level of parallelism executing iterations 3-4 of the i loop. This thread corresponds to

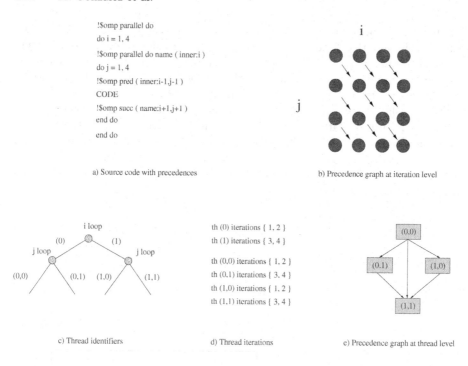

a) Source code with precedences

b) Precedence graph at iteration level

c) Thread identifiers

d) Thread iterations

e) Precedence graph at thread level

Fig. 3. Example of multi level code with precedences.

thread (1,0). The PRED directive inside the thread code will be instantiated four times at runtime, defining four precedences. For the moment, we are interested in the one corresponding to values (i=3) and (j=2). The thread that has to release this precedence is one of the two threads coming out from the parallel branch executing the i-1 (2) iteration in the outer loop, the i loop. Only one of them will execute the iteration j=1. This thread is (0,0). So, in order to determine the synchronizing thread, two iter-to-thread translations have to be performed. One for the outer level of parallelism, the other for the inner one. For each traversed level of parallelism, one translation has to be performed. The runtime has to supply a mechanism to, given the iteration numbers, translate them to the thread identifiers. These thread identifiers plus that of the thread invoking the translation mechanism will be used to determine the memory location used to perform the synchronization operations. In the example the synchronization will be as follows: thread (0,0) signals thread (1,0) and releases the precedence. The tuple created from these identifiers (0, 0, 1, 0) could be used as a sort of coordinates to determine the memory location for the dependence counter. Thread (1,0) should perform an increment on the dependence counter, and wait until the counter reaches zero.

What has been explained for the PRED directive, can be applied to the SUCC directive. Notice that the same coordinates will be obtained if the translation mechanism is applied to the SUCC directive in the thread code for thread (0,0), the one releasing the precedence. That means that the same memory location will be used by thread (0,0) to perform the decrement of the dependence counter. Figure 4 summarizes the basic actions to be taken by the runtime for each PRED and SUCC directive. All iterations appearing in a PRED or SUCC directive have to be translated to thread identifiers. This corresponds to the Translation Phase in Figure 4. After all the thread identifiers are obtained, the memory location accessed to perform the synchronization has to be determined. Then an increment or decrement on that location has to be executed depending on what action is being performed: definition or release of a precedence. This corresponds to the Synchronization Phase.

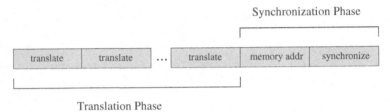

Fig. 4. Runtime actions for precedences.

The translation mechanism has been generally described for the PRED and SUCC but depending on whether the mechanism is invoked for a precedence definition or release, different behaviors have to be specified to deal with some situations. For instance, consider again the previous example of the PRED directive. In that example, for each iteration executed by thread (1) (iterations 3-4 of the outer loop), the inner parallel loop is defined with exactly the same loop parameters. This allows us to predict the output of the translation mechanism for the inner level of parallelism in both iterations. If the loop parameters were different for the two iterations, we could not perform the translation for the inner loop, until it is defined for the translated iteration. Considering the most general situation, it will not be correct to invoke the translation mechanism for an inner loop, until this loop has not been defined for a particular iteration in an outer loop. The runtime has to block the thread invoking the translation mechanism until the iteration currently being translated starts executing. This will ensure that subsequent translations for the inner levels will be performed over the correct loop definitions. Notice that once it is possible to ensure that a translation performed over an inner loop is correct, it is possible that this loop has not been defined yet because the thread executing it has not reached it yet. This will force the runtime to block the translation mechanism for the inner level, until the loop has been defined.

The SUCC directive suffers from the same problem as the PRED directive. But if the same solution is taken (block the thread invoking the translation mechanism), we are degradating the performance of an executing thread free of precedences, but only releasing them. The runtime should distinguish from a translation invoked to release a precedence and a translation invoked to define a precedence. The runtime has to supply an alternative synchronization mechanism (to the dependence counter mechanism) to perform the synchronization in case the required translations to synchronize have not been possible to be performed.

Consider now thread (1,1). One of the precedences that will be defined for this thread involves thread (0,0). Let's focus on the actions that the thread (1,1) has to take to check if the precedence has been released or not. As it was showed, two invocations of the translation mechanism are necessary, one for each traversed parallel region. Once thread (1,1) performs the first translation (for iteration 2 of the i loop), it is possible that for any reason, the thread identified with the translation output (the one executing iteration 2 of the i loop) executes faster, and changes of iteration during or after the first translation is performed, but before the second one has finished or even started. This does not invalid the first translation, but invalids any translation for the inner level, as the definition of the inner loop does not correspond anymore to the one that was required by the translation over the outer level (the definition performed while the execution of iteration 2 in the i loop). Notice that an invalid translation implies that all subsequent translations for inner levels of parallelism are no more necessary to be performed. Any synchronization depending on these translations is assumed to be accomplished. This situation forces the runtime to establish a mechanism to determine whether a translation is valid or not to be performed. The runtime is forced to ensure that any thread executing in the Translation Phase of a precedence, gets correct outputs from the translations, moreover, the runtime has to be able to detect that a translation is not useful anymore because the thread in charge of releasing the precedence has already executed.

Finally, in the example in Figure 3, it is needed to establish a correspondence between the thread identifiers coming out from the translation mechanism to the memory locations to be used as dependence counters. The runtime is in charge of implementing such correspondence.

From the previous explanations, four main aspects have to be faced by the runtime. First, the blocking mechanism for a thread executing in the Translation Phase for the definition of a precedence. Second, never block a thread that is releasing a precedence in case a translation is not possible to be performed. Third, define a mechanism to validate an iter-to-thread translation. Fourth, establish the correspondence between the thread identifiers and the memory locations to be used as dependence counters. The following sections explain how our runtime implementation carries out with this four requirements. Figure 5 shows the code generated by the NanosCompiler for the example in Figure 3. Figure 5.a corresponds to the thread code for the the i loop. Figure 5.b corresponds to the thread code for the j loop.

Blocking Mechanism. First we describe the implementation of the mechanisms that ensure that each time the iter-to-thread translation is invoked, the translation is performed over the correct definition of the involved loop. Remember that there were two reasons that could delay the translation. One corresponds

```
  threadid_i = nthf_rel_whoami()
  CALL nthf_bound_loop(loopdesc_i,loop_actual_iter_i,nth_max,nth_min,
1   4,1,1,01,0,threadid_i)
  DO i = nth_min,nth_max
    CALL nthf_incr_sequence(loopdesc_i,loop_actual_iter_i,i,...)
    CALL p_test_1_01_1_02(loopdesc_i,i,threadid_i,inner_inner_desc,...)
  END DO
  CALL nthf_undefine_loop(loopdesc_i,...)
```

a) Thread code for loop i.

```
  threadid_j = nthf_rel_whoami()
  CALL nthf_define_loop(loopdesc_j,loop_actual_iter_j,nth_max,nth_min,
1 4,1,1,01,1,threadid_i,threadid_j)
  DO j = nth_min,nth_max
    CALL nthf_incr_sequence(loopdesc_j,loop_actual_iter_j,j,1,
1    threadid_i,...)
    prec_s_index_i = nthf_index_to_thread(inner_inner_desc,i - 1,
1    loopdesc_i,id_iter,addr_id_iter,addr_mutex,...)
    prec_s_chunk_inner = nthf_index_to_thread(inner_inner_desc,j - 1,
1    loopdesc_j,id_iter,addr_id_iter,addr_mutex,01,prec_s_index_i)
    CALL nthf_def_prec(inner_inner_desc,id_iter,addr_id_iter,addr_mutex,
1    4,prec_s_index_i,prec_s_chunk_inner,threadid_i,threadid_j,
2    i - 1,j - 1)

    CODE

    prec_t_index_i = nthf_index_to_thread(inner_inner_desc,i + 1,
1    loopdesc_i,id_iter,addr_id_iter,addr_mutex,...)
    prec_t_chunk_inner = nthf_index_to_thread(inner_inner_desc,
1    j + 1,loopdesc_j,id_iter,addr_id_iter,addr_mutex,01,prec_t_index_i)
    CALL nthf_free_prec(inner_inner_desc,loopdesc_j,04,threadid_i,
1    threadid_j,prec_t_index_i,prec_t_chunk_inner)
  END DO
  CALL nthf_undefine_loop(loopdesc_j,1,threadid_i,...)
```

b) Thread code for loop j.

Fig. 5. Code generated for the example in Figure 3.

to the fact that when the translation mechanism is invoked, the target loop is not yet defined. As the loop parameters (lower and upper bounds, step, number of threads executing the loop, ...) are not available, it is impossible to perform the translation. The runtime is forced to block until the loop definition occurs. The runtime routines involved in the detection and management of this situation are nthf_define_loop, nthf_undefine_loop and nthf_index_to_thread. The nthf_define_loop is executed by all the threads participating in the execution of the parallel loop. Its main arguments are a loop descriptor, the loop parameters and a list of thread identifiers indicating from which parallel branch the loop is instantiated (see Figure 5). The routine defines the loop with all the information available. The loop definition is performed by updating a definition flag in the loop descriptor. This routine also updates the content of a vector allocated in the loop descriptor that informs about the state of each of the participating threads. The possible thread states are: EXECUTING, EXECUTED and UNDEFINED. Routine nthf_define_loop sets the state of the thread to EXECUTING. Routine nthf_undefine_loop sets the state to EXECUTED. The use of the thread state will be explained latter. Routine nthf_index_to_thread is in charge of checking the definition flag before starting the translation. In case the flag is not set, the routine blocks the executing thread waiting for the flag to be set and looks for more work to execute.

There was another situation that forced the runtime to block a thread while performing a translation. This situation appeared when the translation was dependent on a previous translation as the target loops of both translations were nested. The nthf_index_to_thread is in charge of managing this situation. Once the loop definition is checked and the translation has been computed, if the thread pointed out by the output of the translation has not started the execution of the translated iteration, the runtime blocks the thread until the required iteration starts executing. This ensures that the next translations will be performed over the appropriate loop instantiations. The loop descriptor appears as argument for the nthf_index_to_thread routine. The loop descriptor contains a vector containing the values of the induction variables of the executing threads. As soon it is checked that the required iteration it is executing, the thread executing the translation is allowed to continue the execution. For this situation the state vector is used, as it is not possible to check the value of a concrete thread induction variable until its state is EXECUTING. Notice that it is possible to compute a translation as soon as the loop parameters are completely defined. It is not necessary that the thread pointed out by the translation is executing.

Buffered Synchronizations. Next we describe how the runtime avoids any unnecessary delay for a thread executing a translation to release a precedence. The situation may appear when the loop parameters are not yet available for a translation. Again routine nthf_index_to_thread is in charge of dealing with this.

An extra argument is given to nthf_index_to_thread routine to indicate that the translation is performed to release a precedence. This avoids any blocking action during its execution. When it is detected that the translation will not be possible to be computed, the runtime reserves an entry in a buffer allocated in the precedence descriptor (notice that the precedence descriptor inner_inner_desc appears as argument of the nthf_index_to_thread routine). The buffer is used to record the list of iterations that could not be translated. In this case the routine nthf_index_to_thread returns a special code indicating which entry has been used. Notice that in Figure 5.b the output of a translation is used as argument for the subsequent translations. This causes that all translations after one that has not been computed, directly access to the same entry in the buffer and records another iteration not translated.

On the other side, when the `nthf_def_prec` routine is invoked to perform the synchronization required to satisfy a precedence, it has to check if the buffer allocated in the precedence descriptor contains any entry with the list of iterations that have been translated. Notice that this iterations appear as arguments in the routine call. In case of an affirmative answer, the synchronization is interrupted and the routine returns assuming that the precedence has been already satisfied. As the size of the buffer is limited, in case it gets full, a thread invoking the translation mechanism to release a precedence should block and wait until the required loop definition is performed.

Translation and Synchronization Validation. Next we describe the runtime mechanism to detect the fact that during the computation of subsequent translations, one of them translates an iteration that has been completely executed, thus invalidating the rest of translations. Routines `nthf_index_to_thread` and `nthf_def_prec` are involved in this. Routine `nthf_index_to_thread` makes the translation as explained before. In case the target loop has not been defined or the requested iteration is not executing, the runtime blocks the translation. Once this is solved, the thread invoking the translation mechanism has to define a dependence to the target loop of the translation. By defining a dependence we mean that the thread has to be binded to the loop thread pointed out by the translation. The way we bind a thread to another thread is implemented by the use of three variables in the invocation of the routines `nthf_index_to_thread` and `nthf_def_prec`. Those variables are: `id_iter`, `addr_id_iter` and `addr_mutex`. Notice that these variables appear as arguments for the invocations to routines `nthf_index_to_thread` and `nthf_def_prec`. Variable `id_iter` contains a sequence number obtained from the loop descriptor at the time it is detected that the translated iteration is being executed. The sequence number is local to the thread pointed out by the translation. This means that the loop descriptor contains different sequence counters as many threads are executing the target loop of the translation. Variable `addr_id_iter` contains an address pointing to the memory location from where the sequence number has been obtained in the loop descriptor. This allows the thread to compare its sequence number with the current sequence number. In case the current sequence number is bigger than the one in the `id_iter` variable, next translations are unnecessary because the precedence has been already satisfied. If the sequence number in the `id_iter` variable is still valid, the next translation can be performed. Each time a translation is accomplished, the thread has to change its dependence and bind to the new loop thread pointed out by the most recently translation. This is a critical action because it has to be guaranteed that during the change, the actual loop thread where the thread is binded is not changing of iteration. To ensure this, variable `addr_mutex` contains the address of a mutex again contained in the loop descriptor and local for each executing loop thread. To perform the new binding the thread locks the mutex. This causes that if the thread pointed out by the translation tries to change the current iteration, it has to wait until the thread executing the binding unlocks the mutex. This is accomplished by the runtime routine nthf_incr_sequence. This routine locks the local thread mutex to perform the increment of the current sequence number and then unlocks the mutex. Once the thread that was making the bind has locked the mutex, asks for a new sequence number and changes the content of the variable addr_id_iter. Then unlocks the mutex and updates the content of variable addr_mutex with the new mutex of the loop thread where it is now binded.

Finally, before executing the **Synchronization Phase** the runtime forces the last check over the last loop thread where the thread checking the precedence is binded. If the loop thread executing the last translated iteration has not changed of iteration, it is correct to increment the dependence counter and wait for the corresponding decrement. If the iteration has been already executed, the precedence has been for sure satisfied. For that reason the three aforementioned variables appear as arguments to the routine nthf_def_prec.

Memory Locations. Finally we present how the runtime determines the memory location to be used for a synchronization. Our implementation uses a list of thread identifiers as indexes for a matrix with as many dimensions as the sum of the levels of parallelism of each of the work–sharing constructs involved in the precedence. In the example in Figure 3, 4 dimensions. Each dimension is sized with the maximum number of available threads. The precedence descriptor contains a base address pointing a memory space used for the synchronizations. The runtime computes an offset according to the list of identifiers. Applying the offset to the base address, a memory location is obtained to for a particular synchronization.

We are conscious that this implementation might lead to a memory allocation much bigger than the real requirements of the application. We are currently optimizing this aspect of the implementation.

4   Conclusions

In this paper we have presented a set of extensions to the OpenMP programming model oriented towards the specification of complex pipelined computations in the context of multilevel parallelism exploitation. Although the majority of the current systems only support the exploitation of single–level parallelism around loops, we believe that multi–level parallelism will play an important role in future systems. When moving towards the exploitation of outer levels of parallelism, it is common to face with the necessity of expressing pipelined computations coming from task graphs (being a task either a chunk of iterations in a parallel loop or a code section). The specification of generic task graphs as well as complex pipelined structures is not an easy task in the framework of OpenMP. In this paper we have presented a simple extension to OpenMP that allows this and relieves the programmer from the burden of defining complex synchronization data structures and the insertion of explicit synchronization actions in the program that make the program difficult to understand and maintain. This work is transparently done by the compiler with the support of the OpenMP runtime library.

The extensions have been implemented in the NanosCompiler and runtime library NthLib. In [4] the authors evaluate the proposal with a synthetic multi-block application on a Origin2000 platform. The results show that when the number of processors is high, exploiting multiple levels of parallelism with thread groups results in better work distribution strategies and thus higher speed-up than both the single level version and the multilevel version without groups. When precedences are taken into consideration, the mechanism proposed at the language level and its implementation in the runtime library are powerful enough to express a variety of scientific applications. The overhead introduced by the mechanisms is compensated by the improved work allocation and results in better performance than the exploitation of a single level of parallelism.

Acknowledgments. This research has been supported by the Ministry of Education of Spain under contract TIC98-511 and the CEPBA (European Center for Parallelism of Barcelona).

References

[1] M. Gonzalez, J. Oliver, X. Martorell, E. Ayguade, J. Labarta and N. Navarro. OpenMP Extensions for Thread Groups and Their Runtime Support. In *Workshop on Languages and Compilers for Parallel Computing*, August 2000.

[2] M. Gonzalez, E. Ayguadé, X. Martorell, J. Labarta, N. Navarro and J. Oliver. Precedence Relations in the OpenMP Programming Model. Second European Workshop on OpenMP, EWOMP 2000 (September 2000).

[3] M. Gonzalez, E. Ayguadé, X. Martorell, J. Labarta, N. Navarro and J. Oliver. NanosCompiler: Supporting Flexible Multilevel Parallelism in OpenMP. Concurrency: Practice and Experience (special issue on OpenMP). Vol.12, no. 12, October 2000.

[4] M. Gonzalez, E. Ayguadé, X. Martorell and J. Labarta. Complex Pipelined Executions in OpenMP Parallel Applications. International Conference on Parallel Processing (ICPP'2001), to appear. September 2001.

[5] X. Martorell, E. Ayguadé, J.I. Navarro, J. Corbalán, M. González and J. Labarta. Thread Fork/join Techniques for Multi-level Parallelism Exploitation in NUMA Multiprocessors. In *13th Int. Conference on Supercomputing ICS'99*, Rhodes (Greece), June 1999.

[6] OpenMP Organization. OpenMP Fortran Application Interface, v. 2.0, www.openmp.org, June 2000.

CableS : Thread Control and Memory System Extensions for Shared Virtual Memory Clusters

Peter Jamieson and Angelos Bilas

Department of Electrical and Computer Engineering
University of Toronto
Toronto, Ontario M5S 3G4, Canada
{jamieson,bilas}@eecg.toronto.edu

Abstract. Clusters of high-end workstations and PCs are currently used in many application domains to perform large-scale computations or as scalable servers for I/O bound tasks. Although clusters have many advantages, their applicability in new areas and especially in areas of commercial applications has been limited. One of the main reasons for this is the fact that clusters do not provide a single system image and thus are hard to program. In this work we address this problem by providing a single cluster image with respect to thread and memory management to programmers. The main limitation of our system is that it does not yet provide file system and networking support across cluster nodes. We implement our system on a 16-processor cluster interconnected with a low-latency, high-bandwidth system area network. We demonstrate the versatility of our system with a wide range of applications. We show that clusters can be used to support applications that have been written for more expensive tightly–coupled systems, with very little effort on the programmer side. Finally, we show that the overhead introduced by the extra functionality of *CableS* affects the parallel section of applications that have been tuned for the shared memory abstraction only in cases where the data placement policy of the system results in improper placement due to operating system limitations in virtual memory mappings granularity.

1 Introduction and Background

Recently there has been progress in building high-performance clusters out of high–end workstations and low-latency, high-bandwidth system area networks (SANs). SANs, used as interconnection networks provide memory–to–memory latencies of under $10\mu s$ and bandwidth in the order of hundreds of MBytes/s . For instance, the cluster we are developing at the University of Toronto uses Myrinet as the interconnection network and currently provides one–way, memory–to–memory latency of about $8\mu s$ and bandwidth of about 120MBytes/s. Similar clusters are being built at many other research institutions.

Moreover, the shared memory abstraction, is used in an increasing number of application areas. Developers have been writing new applications for the shared

R. Eigenmann and M.J. Voss (Eds.): WOMPAT 2001, LNCS 2104, pp. 170–184, 2001.

address space abstraction and legacy applications are being ported to the same abstraction as well. Finally, most vendors are designing both small–scale symmetric multiprocessors (SMPs) and large–scale, hardware cache-coherent distributed shared memory (DSM) systems, targeting both scientific and commercial applications.

Shared memory clusters are an attractive approach to providing affordable and scalable compute cycles and I/O. For this reason, there has recently been a lot of work on designing efficient shared virtual memory (SVM) protocols for such clusters [23,16,26,13]. These protocols take advantage of features provided by SANs, such as low–latencies for short messages and direct remote memory operations with no remote processor intervention [12,10,9], to improve system performance and scalability [16]. Providing a shared memory programming abstraction on clusters has made it easier to run applications that have been written for more traditional, tightly–coupled multiprocessors (both shared bus and distributed shared memory machines).

Recent work [23,16,26] has shown that the performance of SVM clusters is competitive for wide ranges of applications to more traditional, tightly–coupled multiprocessors. For instance, the authors in [16] find that a 64–processor cluster offers, for most SPLASH-2 [24] applications (after a number of optimizations), at least 50% of the performance offered by a 64–processor SGI Origin2000.

However, despite the many advantages of clusters over more traditional, tightly–coupled scalable servers and the competitive performance they offer, their use is not widespread, especially in areas of commercial applications. One of the reasons is that despite the progress on the performance side, it still is a very challenging task to port existing applications or to write new ones for the shared memory programming APIs provided by clusters.

Although these APIs provide sufficient primitives to write parallel programs, they also impose several restrictions: (i) Processes cannot always be created and destroyed on the fly during application execution. (ii) Programmers allocate shared memory only during program initialization and should not free memory until the end of execution. (iii) In most shared memory clusters the synchronization primitives supported are *lock/unlock* and *barrier* primitives. However, more modern APIs support conditional waits as well as other primitives.

These limitations are not very important for large classes of scientific applications that are well structured. However, they pose important obstacles for using clusters in areas of applications that exhibit a more dynamic behavior, such as commercially–oriented applications. Thus, in many areas, traditional shared memory multiprocessors are used because of their ability to support legacy applications with no or very few changes. Also, development of new applications usually occurs on these architectures, since they provide APIs that pose fewer restrictions to the programmer. In essence, current clusters that support shared memory provide a very limited single system image to the programmer with respect to process, memory management, and synchronization.

In order to overcome the above limitations and to provide a more complete and functional single cluster image to the programmer, we design and implement a *pthreads* interface on top of our cluster. This allows existing *pthreads* programs to run on our system with minor modifications. Programs can dynamically create and destroy threads on our system, allocate global shared memory throughout execution, and use all synchronization primitives specified by the *pthreads* API. More specifically, our system, *CableS* (**C**luster enable**d** thread**S**), provides:

Support for dynamic node and thread management: *CableS* allows the application to dynamically create threads at any point during execution. Currently, new threads are allocated to nodes with a simple, round–robin policy. On the fly, the system performs all the necessary initialization to support the *pthreads* API.

Support for dynamic memory management: *CableS* addresses a number of issues with respect to memory management. (a) It provides all necessary mechanisms to support different memory placement policies. Currently, *CableS* supports first touch placement, but can be extended to support other policies as well. (b) It provides the ability to allocate global, shared memory dynamically at any time during program execution. (c) It deals with static global variables in a transparent way.

Support for modern synchronization primitives: *CableS* supports the conditional wait primitives.

The main limitation of *CableS* is that, although it provides a single system image with respect to thread management, memory management, and synchronization support, it does not yet include file system and networking support across cluster nodes.

We demonstrate the viability of our approach and the versatility of our system by using a wide range of applications: (a) We run existing *pthreads* applications with minor modifications. (b) We use a public–domain OpenMP compiler, OdinMP [7], that translates OpenMP programs to *pthreads* programs for shared memory multiprocessors and run the translated OpenMP programs on our system. Our system supports the OpenMP programs with no modifications to the OpenMP source and minor modifications to the *pthreads* sources. (c) We provide an implementation of the M4 macros for *pthreads* and we run some SPLASH-2 applications.

We also show that the overhead introduced by the extra functionality affects the parallel section of applications that have been tuned for the shared memory abstraction only in cases where the data placement policy of the system results in improper placement due to operating system limitations in virtual memory mappings granularity. In the SPLASH-2 applications most overhead is introduced during application initialization and termination, whereas the execution time of the parallel section is only affected by the data placement, currently determined by our first touch policy.

The paper is organized in the following sections. Section 2 introduces our platform and current architectural considerations for modern clusters. Section 3 describes the design of *CableS*. Section 4 presents our experimental results. Section 5 presents related work and Section 6 discusses our high level conclusions.

2   Modern Clusters and System Area Networks

Nodes in modern clusters are usually interconnected with low–latency, high–bandwidth SANs that support user–level access to network resources [12,9,5]. By allowing users to directly access the network without operating system intervention, these systems dramatically reduce latencies compared to traditional TCP/IP based local area networks. Moreover, to further reduce latencies in SANs, direct memory operations are usually supported; reads and writes to remote memory are performed without remote processor intervention.

This mechanism provides fast access to remote memory within a cluster. However, there are a number of limitations associated with these operations and with modern SANs in general: (i) Connection establishment requires mapping of remote memory locally, which is an expensive operation. This operation usually sets up some form of page table on the network interface card (NIC) that allows it to access remote memory. (ii) The amount of remote memory that can be mapped is limited. (iii) To provide asynchronous communication primitives, most communication layers use a notification mechanism. Notifications use the interrupt mechanism provided by the operating system and are usually very expensive compared to direct remote operations.

The specific system we use consists of a cluster of 8 2-way PentiumPro SMP nodes interconnected with a Myrinet network. Each SMP is running WindowsNT. The nodes in the system are connected with a low–latency, high–bandwidth Myrinet SAN [5]. The software infrastructure in the system includes a custom communication layer and a highly optimized SVM system.

The communication layer we use on top of Myrinet is a user-level communication layer, Virtual Memory Mapped Communication (VMMC) [2,9]. VMMC provides both explicit, direct remote memory operations (reads and writes) and notification–based send primitives.

The SVM protocol used is GeNIMA [16], which is a home-based, page-level SVM protocol. The consistency model in the protocol is Release Consistency [11]. GeNIMA provides an API based on the M4 macros, which are extensively used for writing shared memory applications in the scientific computing community.

3   System Design

CableS supports a full *pthreads* (POSIX threads IEEE POSIX 1003.1 [1]) API which enables legacy shared memory applications written for traditional, tightly coupled, hardware shared memory systems to run on shared memory clusters. Within the *pthreads* API, *CableS* addresses the following issues: (i) It supports the *pthreads* API. (ii) It provides support for dynamic global memory management.

3.1   Thread Management

Thread and node management is the core of the API in which a thread can be created and administrated. The API also includes mechanisms to kill threads, cancel threads, and store thread private data.

In a distributed environment, threads of execution need to be started and administered on remote systems. For this purpose, *CableS* needs to maintain and manage global state that stores location and resource information about each thread in an application. Thus, *CableS* uses per application global state, called the application control block (ACB). This state is updated by all nodes in the system via direct remote operations as well as notification handlers.

CableS' communication library, VMMC, provides communication primitives that allow one node to perform reads and writes to another node's memory without interrupting the remote processor. *CableS* maintains the most up to date system information on the first node where the application starts (master node). To ensure consistency of the ACBs, updates are performed either by the master node through remote handler invocations, or by node update regions in which the system guarantees that the node is the exclusive writer.

The thread management component of the *pthreads* library is hinged around thread creation. Thread creation in *CableS* involves one of three possible cases: (i) Create a thread on the local node. (ii) Create a thread on a remote node that is not used by this application. This operation is called attaching a remote node to the application. (iii) Create a thread on an already attached remote node.

Local thread creation is equivalent to a call to the local operating system to create a thread. *CableS* creates remote threads by a combination of direct memory operations and remote handler invocation; thus, remote thread creation is expected to be a fairly expensive operation.

When *CableS* needs to attach a new node to the application, the master node M creates a remote process on the new node N. Node N, starts executing the initialization sequence and performs all necessary mappings for the global shared memory that is already allocated on M. N then retrieves global state information from M including shared memory mappings and sends an initialization acknowledgment back to M. M broadcasts to all other nodes in the system that N exists and that they can establish their mappings with N. At the end of this phase, node N has been introduced into the system and can be used for remote thread creations.

The remaining thread management operations involve mostly state management, mainly, through direct reads and writes to global state in the ACB.

3.2   Synchronization Support

The *pthreads* API provides two synchronization constructs *mutexes* and conditions. Current SVM APIs that mostly target compute-bound parallel applications provide two other synchronization primitives, locks and barriers.

Since mutexes and locks are very similar, we use the underlying SVM lock mechanism to provide mutexes in the *pthreads* API. For performance reasons, locks are implemented in SVM as spin locks, and we maintain this implementation in *CableS*.

The *pthreads* condition is a synchronization construct in which a thread waits until another thread sends a signal. As with mutexes, conditions can be implemented either by spinning on a flag or by suspending the thread on an operating system event. Although implementations that use spinning consume processor cycles, they are more common in parallel systems to reduce wake–up latency. For this reason, our first implementation of *pthreads* conditional wait primitives uses spinning.

Global synchronization (barriers) can be implemented in *pthreads* with mutexes (or conditions). However, to support legacy parallel applications efficiently we extend the *pthreads* synchronization to support a barrier operation.

3.3   Memory Subsystem

Current System Limitations. Modern SANs that support direct remote memory operations, such as remote read and write operations require some form of memory registration to avoid remote processor intervention and interrupts. In these mechanisms, a node maps one or more regions of remote memory to the local network interface card (NIC) and it performs direct operations on these regions without requiring processor intervention on the remote side. This mapping operation is called registration and usually requires work at both the sending as well as the receiving NIC. SVM systems on clusters interconnected with SANs take advantage of these features to reduce the overhead associated with propagation and obtaining updates of shared data [23,16]. For this purpose, they perform all necessary registration operations at initialization time. This results in a number of limitations:

All shared memory has to be allocated at initialization time, since all memory registration operations happen at initialization.

In most SANs today [8,12,9] there is a limit (a) on the number of memory regions and (b) on the total amount of memory that can be registered on the NIC. Each page in the working set of each process should be placed, for performance reasons, on the node where the process runs. The registration limitations conflict with this requirement.

On today's systems there exist three possible solutions to this problem: (i) Shared pages could be grouped in regions and mapped together to solve registration limitations. In this case, pages in the working set of a process will have their primary copies in remote nodes resulting in excessive network traffic and performance degradation. (ii) Place the primary copies of pages in the working set on the node where the process runs. In this way the registration limitations may be violated since there will be a large number of non-contiguous memory regions that have to be registered. (iii) The many non-contiguous regions could be registered in one operation, including the gaps between regions. However, this results in registering essentially all the shared address space. This is not feasible due to the total amount of memory which can be registered.

In most SVM systems today, global static variables are not included in the shared address space. These are global variables that are declared statically in the user program. In the threads programming model, these variables are visible to all threads; however, this is not true in most SVM systems. The compiler/linker automatically allocates these variables to a designated part of the virtual address space. Since this part of the address space is not under the control of the shared memory protocol, static global variables can not be shared across nodes. This imposes additional challenges in the process of porting existing shared memory applications to clusters.

Effects on System API. The above limitations impact the API provided to users in many ways. To support a global shared address space on a cluster without hardware support, most systems today perform the following steps:

1. Start and initialize all the nodes that will be used to run the application at the same time.
2. Allocate a region of the virtual address space on each node. This step is usually fairly simple, since it involves using a system call to reserve part of the application virtual address space.
3. Determine which node will maintain the primary copy of each portion of the shared memory.
4. Establish communication–layer mappings between the primary copy of this region and the same region on all other nodes. These mappings are established by filling in the necessary information in the NIC page table.
5. Provide the application on each node with a convenient interface to the shared address space. This must consider the current restrictions on the usage of the shared address space: applications cannot use static global variables, nor allocate/deallocate shared memory after thread creation.

Proposed Solution. Existing systems deal with these issues by imposing API limitations that make it easy to avoid the related problems. The result is inflexible systems that are not easy to program. *CableS* deals with most of the issues above as follows.

Shared memory allocation and registration: Initially, one contiguous part of the physical address space in each node is used to hold the primary copies of shared pages that will be allocated to this node. This part of the physical address space is always pinned (can't swap out of RAM), since it will be accessed remotely by other nodes. The primary copies are mapped twice to the virtual address space of the process. One mapping is to a contiguous part of the virtual address space and is used only by the protocol . The second mapping is used by the application to access the shared data. For this mapping, the home pages are divided in groups of fixed size (in the current system 64 K-Bytes) and are mapped to arbitrary locations in the virtual address space of the process. It is important to note that these locations are not necessarily contiguous.

As the application requires more shared memory, it first allocates a region in the global virtual address space. Then, it determines which node will hold the primary copies of these pages according to some placement policy (currently first touch).

When a home page is touched: (a) The home node extends the home pages section and registers the additional pages with the NIC. Then, it maps the virtual memory region to the newly allocated home pages. As the primary copies of shared pages are placed in different nodes, the home pages portion of the physical address space is mapped to non-contiguous regions of the shared virtual address space in the home node. (b) Every other node in the system, registers the newly allocated virtual memory region with the NIC so that each node can fetch updates from the primary copies and rely on the OS to allocate arbitrary physical frames for these pages.

First touch policy: Implementing a first touch policy requires that the system delays binding of virtual addresses until the region is first read or written. *CableS* maintains information about each memory segment allocated in the global directory. During execution, when a node touches the segment, it uses the global directory to identify if the segment has been touched by anyone else. If it has, then the segment is registered with the NIC and is mapped to the corresponding region on the home node. If this is the first touch to the region, then the node becomes the home by updating the global information and by appropriately mapping the physical pages to its shared virtual address space so that the application can use it. Synchronization of the global information and ordering simultaneous accesses to a newly allocated region is facilitated through system locks.

Global static variables: *CableS* deals with global static variables in a transparent way. In current SVM systems and related APIs, such as M4, global static variables can only be pointers, allocated explicitly by the programmer. Explicit allocation greatly simplifies management of these variables, since the system can allocate them in designated areas of the shared address space [1]. However, this imposes a burden on the programmer, since they need to allocate each global variable explicitly. Moreover, these variables greatly hinder porting of existing *pthreads* applications to clusters.

CableS uses a type quantifier to allocate these global variables in a special area within the executable image. At application initialization, the first node in the system becomes the primary copy for this region. All necessary mappings are established to other nodes as they are attached to the application. Thus, static global variables of arbitrary types can be shared among system nodes.

---

[1] This assumes that the designated part of the virtual address space for static variables is the same in all processes. This is true in most systems today (or can be easily enforced).

3.4   Summary

CableS provides a shared memory programming model that is very similar to a *pthreads* programming model for tightly–coupled shared memory multiprocessors, such as SMPs and hardware DSMs. To run any *pthreads* program on *CableS*, the following modifications are required:

1. Determine if the program will perform correctly under a Release Consistency memory model.
2. Add the *pthread_start* and *pthread_end* library calls.
3. Prefix all static variables that will be globally shared with the *GLOBAL* identifier.
4. (OPTIONAL) Optimize the code so that threads touch data they require.
5. Link with *CableS*' library.

4   Results

In this section we present three types of results: (i) We demonstrate that legacy *pthreads* programs written for traditional hardware shared memory multiprocessors can run with minor modifications on *CableS*. (ii) As an extension of (i) we show that OpenMP programs can be run by translating them to *pthreads* programs by using a public domain OpenMP compiler, OdinMP [7]. (iii) We provide an implementation of the M4 macros for *pthreads* and run a subset of the SPLASH-2 applications.

4.1   Legacy *pthreads* Programs

We use the five simple steps, outlined above, to convert three publicly available *pthreads* programs for *CableS*. The programs are: (i) Prime number (PN), which, as indicated by its name, computes all prime numbers in a user specified range. (ii) Producer–consumer (PC), a producer–consumer program which runs with two threads. (iii) Pipe (PIPE), which creates a threaded pipeline where each element stage consists of a calculation. Next, we use OdinMP to compile to *pthreads* three SPLASH-2 applications that have been written for OpenMP: FFT, LU, and OCEAN.

Table 1 shows the *pthreads* programs which were run on *CableS*, and the *pthreads* calls each of the programs make, along with the average execution time of each *pthreads* functions. Note that PC only uses two threads; therefore, this program runs on only one node.

Performance-wise, PC shows the approximate cost of local API operations. PN, PIPE, and the OpenMP programs provide an indication of the average execution time of remote operations in *CableS*. We see that local operations are about three orders of magnitude faster than remote operations. With respect to synchronization operations, conditional waits and mutex lock operations include the cost of communication and the application wait time. For example, in PIPE the condition is used by each stage of the pipe to wait for work. Therefore, a stage

Table 1. Shows pthread programs with their respective pthread function calls and execution times (in ms) for the basic API operations.

PROGRAM	C	J	L	Co	Ca	K	G	Cr	Lo	Un	Wa	Si	Br	Sp
PN	•	•	•	•	•		•	2254	23	2	6154	-	1	15677
PC	•	•	•	•			•	1.1	0.05	0.005	17	0.042	-	-
PIPE	•		•	•			•	1008	52	3	527	12	-	11249
OMP_FFT	•		•	•		•	•	1235	54	0.52	1382	0.146	1.1	12302
OMP_LU	•		•	•		•	•	1247	133	1	327	0.134	0.401	12412
OMP_OCEAN	•		•	•		•	•	1312	49	2	494	0.293	0.606	14222

LEGEND: **C** = pthread_create, **J** = pthread_join, **L** = mutexes, **Co** = conditions, **Ca** = thread cancel, **K** = thread specific information, **G** = program uses static global variables **Cr** = create, **Lo** = mutex locks, **Un** = mutex unlock, **Wa** = condition wait, **Si** = condition signal, **Br** = condition broadcast, **Sp** = spawn time.

in the pipe is dependent on the execution time of the previous stage. Condition signals and broadcasts are much faster since these involve sending only small messages to activate threads in remote nodes.

Table 2. Speedups for the three SPLASH-2 OpenMP programs

PROGRAM	4 processors	8 processors	16 processors
FFT	1.61	2.05	2.44
LU	3.17	3.71	7.10
OCEAN	1.33	1.43	1.92

Table 2 shows the speedups of the three OpenMP SPLASH-2 applications. We do not directly compare these results with our M4 results since OdinMP introduces overheads when translating OpenMP programs into *pthreads*. Also, we have not modified the resulting *pthreads* programs for optimal data placement.

4.2   SPLASH-2 Applications

To investigate the overhead *CableS* introduces in applications which have been tuned for the shared memory abstraction, we provided an implementation of the M4 macros on *CableS* and run a subset of the SPLASH-2 applications on two configurations: The original, optimized SVM system that we started from, GeN-IMA [16], and *CableS*. In *CableS* we use the earlier introduced *pthreads* barriers, as opposed to a mutex-based implementation of barriers which only uses native *pthreads* calls. The motivation behind this extension is that the *pthreads* was not designed for parallel applications which frequently require global synchroniza-

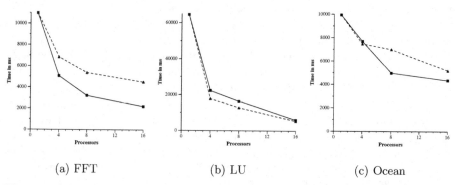

(a) FFT   (b) LU   (c) Ocean

Fig. 1. SPLASH-2 M4 vs M4-pthread executions with 1, 4, 8 and 16processors. Solid line is the M4 executions, and dashed line isM4-pthread executions

tion. For a better comparison, specific knowledge provided by the SPLASH-2 applications about global synchronization should be exploited in both systems.

The applications we use are: FFT [4,24,25], LU [24,25], and OCEAN [6,22, 17]. Their common characteristic is that they are optimized to be single-writer applications; meaning, a given word of data is written only by the processor to which it is assigned. Given appropriate data structures, these applications are single-writer at page granularity as well, and pages can be allocated among nodes such that writes to shared data are almost all local. The applications have different inherent and induced communication patterns [24,14], which affect their performance and the impact on nodes.

Figure 1 shows the execution times of each application in both system configurations for 1,4,8, and 16 processors. We see that the current implementation of the first touch placement in *CableS*. Although this implementation results in similar speedup curves, it increases the absolute execution time in applications where the 64-KByte mapping granularity imposed by the operating system results in improper data placement. In these applications, although the granularity of sharing is still one page (4 KBytes), data is placed in nodes in chunks of 64 KBytes. This may result in additional diff computations, with more expensive page faults and synchronization.

FFT incurs higher data wait time on the first node as shown in Figures 2-a and 2-b; this leads to higher barrier synchronization time. Although the main data structures in FFT are placed properly by *CableS*, there are smaller data structures that reside on node 0. The original system, however, allocated these data structures in a round robin fashion. The reason for this difference is that *CableS* performs first touch allocation, and since node 0 initializes these data structures, all other nodes fetch from node 0. Each node in the system fetches about 5200 pages while node 0 only fetches 3600. Given that in the current VMMC implementation the incoming path has priority over the outgoing path, remote requests for shared pages create contention in the I/O and memory bus of node 0. This means page fetches incur higher delays resulting in high data wait times. Finally, the additional traffic on the memory bus of node 0 increases memory overhead and affects compute time. One way to address this issue would be to distribute all application data structures properly.

Given the large granularity in LU (Figures 2-c and 2-d), the 64-KByte mapping granularity is not an issue. In fact, the performance of the parallel section is almost identical between the two configurations.

OCEAN (Figures 2-e–2-f) incurs higher overheads due to the higher granularity of data placement. The 64-Kbyte chunk size is a major data placement issue since 16 contiguous pages must have the same home node. OCEAN does not have large contiguous regions of data and suffers from misplaced pages. This causes contention within the network which increases synchronization overheads by about 139the average for locks and 106

5   Related Work

This work provides a *pthreads* API for a cluster interconnected with a SAN similar to DSM-Threads [21]. This work targets the implementation of a *pthreads* API on clusters of workstations. *CableS*, however, deals thoroughly with outstanding memory issues. The *pthreads* standard is defined in [1]. Most other related work in the area has focused on the following four directions:

To improve the performance of SVM on clusters with SANs. There is a large body of work in this category [23,18,16,26] . Our work relies on the experiences gained in this area and builds upon it to extending the functionality provided by today's clusters.

To provide OpenMP implementations for clusters. Relatively, little work has been done in this area. The authors in [19] provide an OpenMP implementation based on TreadMarks. They convert OpenMP directly into TreadMark system calls. They then compare the OpenMP programs to the native versions of the same applications.

To provide a *pthreads* interface on hardware shared memory multiprocessors, either shared–bus or distributed shared memory. Most hardware shared memory system and operating system vendors provide a *pthreads* interface to applications [20]. In many systems, this is the preferred API for multithreaded applications due to the portability advantages.

To provide a single system image on top of clusters. These projects focus on providing a distributed operating system on top of clusters. The focus is on providing an operating system that can manage all aspects of a cluster in multitasking environments and not on parallel applications. Also, the authors in [3] provide a Java Virtual Machine on top of clusters. This work focuses on Java applications and uses the extra layer of the JVM to provide a single cluster image. Our work is at a lower layer. For instance, a JVM written for the *pthreads* API, such as Kaffe [15] could be ported to our system.

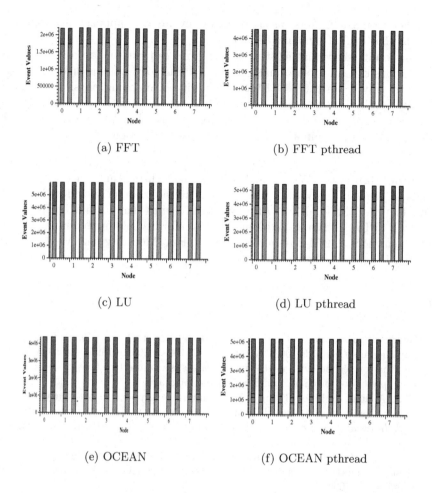

(a) FFT

(b) FFT pthread

(c) LU

(d) LU pthread

(e) OCEAN

(f) OCEAN pthread

Fig. 2. SPLASH-2 M4 and M4-pthread execution breakdowns on16 processors. For (a), (b), (c), and (d) the lower boxes show compute time, the middle boxes show data time, and the top boxes show barrier time. For (e) and (f) the lower boxes show compute time, the next boxes show data time, the 2nd highest boxes show lock time, and finally, the top boxes show barrier time

6   Conclusion

In this work we design and implement a system that provides a single system image for SVM clusters. Our system supports the *pthreads* API and within this API, provides dynamic thread and memory management as well as all synchronization primitives. We show that this system is able to support *pthreads* applications written for more tightly–coupled, hardware shared memory multiprocessors. We use a wide suite of programs to demonstrate the viability of our approach to make clusters easier to use in new areas of applications, especially in areas that exhibit dynamic behavior.

Our results show that existing applications can run on top of *CableS*, and applications tuned for performance on shared memory systems incur additional overhead only when the 64-KByte granularity of mapping physical to virtual memory results in a deviation from the first touch allocation and an improper data placement. The rest of the overhead introduced by *CableS* is limited to the initialization and termination sections of these applications.

Acknowledgments. We would like to thank Jeffrey Tang for his help with fixing problems in the VMMC firmware and Reza Azimi for providing help with extending the VMMC driver. Also, we would like to thank Anna Thelin for her OpenMP SPLASH-2 code, and Mats Brorsson for his help in obtaining OdinMP and OpenMP SPLASH-2 resources. Finally, Paul McHardy and Alexis Armour for their insights on the paper.

References

1. International standard iso/iec 9945-1: 1996 (e) ieee std 1003.1, 1996 edition (incorporating ansi/ieee stds 1003.1-1990, 1003.1b-1993, 1003.1c-1995, and 1003.1i-1995) information technology – portable operating system interface (posix) – part 1: System application program interface (api) [c language].
2. J. S. A.Bilas, C Liao. Using network interface support to avoid asynchronous protocol processing in shared virtual memory systems. In *Proceedings of the The 26th International Symposium on Computer Architecture*, Atlanta, Georgia, May 1998.
3. Y. Aridor, M. Factor, A. Teperman, T. Eilam, and A. Schuster. A high performance cluster jvm presenting a pure single system image. In *ACM Java Grande 2000 Conference*, 2000.
4. D. H. Bailey. FFTs in External or Hierarchical Memories. *Journal of Supercomputing*, 4:23–25, 1990.
5. N. J. Boden, D. Cohen, R. E. Felderman, A. E. Kulawik, C. L. Seitz, J. N. Seizovic, and W.-K. Su. Myrinet: A gigabit-per-second local area network. *IEEE Micro*, 15(1):29–36, Feb. 1995.
6. A. Brandt. Multi-level adaptive solutions to boundary-value problems. *Mathematics of Computation*, 31(138):333–390, April 1977.
7. C. Brunschen and M. Brorsson. Odinmp/ccp - a portable implementation of openmp for c. *The 1st European Workshop on OpenMP*, 1999.
8. D. Cohen, G. G. Finn, R. Felderman, and A. DeSchon. The use of message-based multicomputer components to construct gigabit networks. *ACM Computer Communication Review*, 23(3):32–44, July 1993.
9. C. Dubnicki, A. Bilas, Y. Chen, S. Damianakis, and K. Li. VMMC-2: efficient support for reliable, connection-oriented communication. In *Proceedings of Hot Interconnects*, Aug. 1997.
10. D. Dunning and G. Regnier. The Virtual Interface Architecture. In *Proceedings of Hot Interconnects V Symposium*, Stanford, Aug. 1997.

11. K. Gharachorloo, D. Lenoski, and et al. Memory consistency and event ordering in scalable shared-memory multiprocessors. In *In 17th International Symposium on Computer Architecture*, pages 15–26, May 1990.
12. R. Gillett, M. Collins, and D. Pimm. Overview of network memory channel for PCI. In *Proceedings of the IEEE Spring COMPCON '96*, Feb. 1996.
13. L. Iftode, C. Dubnicki, E. W. Felten, and K. Li. Improving release-consistent shared virtual memory using automatic update. In *The 2nd IEEE Symposium on High-Performance Computer Architecture*, Feb. 1996.
14. L. Iftode, J. P. Singh, and K. Li. Understanding application performance on shared virtual memory. In *Proceedings of the 23rd International Symposium on Computer Architecture (ISCA)*, May 1996.
15. T. T. Inc. Wherever you want to run java, kaffe is there.
16. D. Jiang, B. Cokelley, X. Yu, A. Bilas, and J. P. Singh. Applicaiton scaling under shared virtual memory on a cluster of smps. In *The 13th ACM International Conference on Supercomputing (ICS'99)*, June 1999.
17. D. Jiang, H. Shan, and J. P. Singh. Application restructuring and performance portability across shared virtual memory and hardware-coherent multiprocessors. In *Proceedings of the 6th ACM Symposium on Principles and Practice of Parallel Programming*, June 1997.
18. P. Keleher, A. Cox, S. Dwarkadas, and W. Zwaenepoel. Treadmarks: Distributed shared memory on standard workstations and operating systems. In *Proceedings of the Winter USENIX Conference*, pages 115–132, Jan. 1994.
19. H. Lu, Y. C. Hu, and W. Zwaenepoel. Openmp on networks of workstations. In *Proceedings Supercomputing*, 1998.
20. F. Mueller. A library implementation of posix threads under unix. In *Proceedings of the USENIX Conference*, pages 29–41, Jan. 1993.
21. F. Mueller. Distributed shared-memory threads: Dsm threads. *Workshop on Run-Time systems for Parallel Programming*, pages 31–40, April 1997.
22. J. P. Singh and J. L. Hennessy. Finding and exploiting parallelism in an ocean simulation program: Experiences, results, implications. *Journal of Parallel and Distributed Computing*, 15(1):27–48, May 1992.
23. R. Stets, S. Dwarkadas, N. Hardavellas, G. Hunt, L.Kontothanassis, S. Parthasarathy, and M. Scott. Cashmere-2L: Software Coherent Shared Memory on a Clustered Remote-Write Network. In *Proc. of the 16th ACM Symp. on Operating Systems Principles (SOSP-16)*, Oct. 1997.
24. S. Woo, M. Ohara, E. Torrie, J. P. Singh, and A. Gupta. Methodological considerations and characterization of the SPLASH-2 parallel application suite. In *Proceedings of the 23rd International Symposium on Computer Architecture (ISCA)*, May 1995.
25. S. C. Woo, J. P. Singh, and J. L. Hennessy. The performance advantages of integrating message-passing in cache-coherent multiprocessors. In *Proceedings of Architectural Support For Programming Languages and Operating Systems*, 1994.
26. Y. Zhou, L. Iftode, and K. Li. Performance evaluation of two home-based lazy release consistency protocols for shared virtual memory systems. In *Proceedings of the Operating Systems Design and Implementation Symposium*, Oct. 1996.

Author Index

Lecture Notes in Computer Science

For information about Vols. 1–2025
please contact your bookseller or Springer-Verlag